SpringerBriefs in Electrical and Computer Engineering

More information about this series at http://www.springer.com/series/10059

Igor Vujović

Multiresolution Approach to Processing Images for Different Applications

Interaction of Lower Processing with Higher Vision

Springer

Igor Vujović
Faculty of Maritime Studies,
Department of Marine Electrical
Engineering and Information
Technologies
University of Split
Split
Croatia

ISSN 2191-8112 ISSN 2191-8120 (electronic)
SpringerBriefs in Electrical and Computer Engineering
ISBN 978-3-319-14456-6 ISBN 978-3-319-14457-3 (eBook)
DOI 10.1007/978-3-319-14457-3

Library of Congress Control Number: 2014958573

Springer Cham Heidelberg New York Dordrecht London

Printed on acid-free paper

Springer International Publishing AG Switzerland is part of Springer Science+Business Media (www.springer.com)

Preface

When I was invited to deliver a speech on the presented topic at the 8th International Conference on Advanced Computational Engineering and Experiment (ACE-X 2015) in Paris, I had my doubts as to whether and how I could make this topic interesting to the entire audience, rather than only to a couple of scientists. In the times of budget cuts and reduced investment in science (in Croatia at least), what could I offer that would be interesting to all listeners? The presentation should be interesting, informative, and understandable to the audience consisting of specialists from a variety of technical research fields. That excludes mathematical considerations, equations, theorems, and the like. My benchmark was a comment of my coworker who said that I do not need to present and compare state-of-the-art narrow topics, but offer a broader perspective on our research. The audience would like to hear what we are doing, what we have done, and what the options for further research are. The conclusion was that an overview of the current and previous research of our team could give a wider, interesting perspective to the entire audience.

The book covers several research fields studied by me and my coworkers from the Faculty of Maritime Studies, University of Split. We are making headway in all these areas and expect to get some new results soon. Some results presented here are preliminaries and need to be confirmed in the future. Research fields include:

1. video surveillance,
2. biomedical applications,
3. improved communications through teleoperation, telemedicine, animation, augmented/virtual reality, and robot vision,
4. monitoring of the condition of a ship's systems and image quality control.

The field of video surveillance includes the impact of weather conditions on the system's performance, security and safety applications, traffic monitoring and control, outdoor, indoor, and similar applications. Since image processing is of key importance here, this research filed deals not only with all other aspects of low-level image processing, but high-level vision applications as well. This research field is not surprising since I am a maritime faculty employee and we, among other things, also study maritime transportation.

The research field of biomedical applications was mostly based on image processing until 2014. It included the diagnosis of occupational asbestosis by X-ray images and visualization of anomalies in medical images. This part of research was conducted in cooperation with my colleges from the Faculty of Electrical Engineering and Computing and School of Medicine at our University. We started to analyze electromyographic (EMG) signals in 2014, which are not images, but recordings of the brain's electrical signals. This research field is of interest to maritime experts because EMG can help determine the influence of long contracts on the mental health of seafarers.

The most interesting research field is improved communications due to the strong phrases used, but it is harder for experimentation due to funding difficulties. In this research, we collaborated with faculties of electrical engineering and computing.

Although there is no obvious connection between the research fields of condition monitoring and image quality control, we found it in materials as parts of maintenance, degradation research, and aging studies. We are monitoring ship vibrations to determine their possible impact on various ship systems like steering systems. Vibrations can result in false readings and could have considerable impact on the durability of different elements of ship systems.

Integral transforms in signal processing are the background to all applications. I intentionally started with research fields, because signal processing is not interesting to the wider audience. The actual task my team is working on is signal processing. The character of the signals—depends on the application.

In this book, I will try to present some aspects of each research field.

I would like to thank prof. Fabiana Rodrigues Leta for inviting me on such a wonderful voyage of exploration of the interesting and new world of science. It is evident from this book that many applications are interconnected through similar or even same algorithms, models, background, math, or even line of thinking.

I would like to thank prof. Andreas Oechsner for his assistance with Springer.

I would also like to thank my colleague prof. Ivica Kuzmanić, who recommended me to prof. Leta, and Joško Šoda (senior lecturer and research associate) whose constructive comments and advice about the speech and the presentation were most helpful.

Acknowledgments

I would like to thank prof. Fabiana Rodrigues Leta for inviting me to deliver a speech on this subject. Without it, this book would never happen.

I would like to thank prof. Andreas Oechsner for assistance with Springer. He is the main organizer of the conference series—International Conference on Advanced Computational Engineering and Experiment (ACE-X).

I would also like to thank my colleague prof. Ivica Kuzmanić, who recommended me to prof. Leta and Joško Šoda (senior lecturer and research associate) whose constructive comments and advice on the speech and the presentation were most helpful.

I would also like to thank the reviewers who have shown understanding while assessing this work.

The equipment used in this research was partly obtained by different projects at the University of Split, Faculty of Maritime Studies, and particularly by the scientific research project no. 250-2502209-2364 and the international research project "The possibilities of reducing pollutant emissions from ships in the Montenegrin and Croatian Adriatic implementing Annex VI of MARPOL Convention" supported by the Ministry of Science, Education and Sport of the Republic of Croatia.

Contents

Abbreviations

2D-DWT	Two-dimensional Discrete Wavelet Transform
ANN	Artificial Neural Network
CAD	Computer-aided Diagnostics
CCD	Charge Coupled Device
CR	Compression Ratio
DOF	Degree of Freedom
DWT	Discrete Wavelet Transform
EDF	European Data Format
EM	Energy Measure
EMD	Empirical Mode Decomposition
EMG	Electromyographic
FT	Fourier Transform
HF	High Frequency
IWT	Inverse Wavelet Transform
LF	Low Frequency
LWT	Lifting Wavelet Transform
NNC	Neural Network Classifier
PCC	Percentage of Correct Classifications
ROI	Region of Interest
SAR	Synthetic Aperture Radar
SGW	Second Generation Wavelets
SHM	Structural Health Monitoring
STFT	Short-term Fourier transform
USCGC	United States Coast Guard Cutter
WE	Wavelet Energy
WT	Wavelet Transform

Abstract

Nowadays, computers are expected to perform complex tasks involving the processing of huge amounts of data. People are often unaware that such intensive operations are required in computer vision tasks. Different phenomena related to pixel/voxel or global size can cause failure of higher vision applications. Such phenomena include illumination variations, noise, camera jitter, shadows, visibility, weather conditions, etc. This paper analyzes the influence of lower vision tasks (e.g., denoising or thresholding) on higher vision tasks (e.g., motion segmentation or product quality). The interaction between lower and higher vision is illustrated with examples of visual quality control and advanced visualization in marine communications used to decrease the stress felt by seafarers due to their separation from their families.

Chapter 1
Introduction

Abstract Application of the image processing is popular due to the fact that cameras are widely used sensors in automated systems nowadays. Motivation for the book is found in the research examples and explained in the introduction. This chapter presents an introduction to the book contents. Organization of the book is presented.

Keywords Image processing applications · Data visualization · Wavelets · Multiresolution approach

Due to rapid technological development, and especially the development of sensor technology and computational power, a huge number of automated systems is being developed and installed. Their purpose is to make our everyday lives easier. Cameras are some of the most widely used sensors in automated systems. Fields like medicine and engineering are almost unimaginable without cameras. For example, nowadays, surgeries are performed using cameras, allowing the surgeon either to see what he is doing, or to transmit information over the Internet, or as a diagnostic tooldiagnostic tool. Surveillance applications are also very important and widely used, for example in traffic surveillance or to secure an area or a building, as well as in computer vision applications in the automated production industry and quality control systems. Since cameras obviously necessitate the processing of huge amounts of information, image processing is a rapidly growing field. The main purpose of image processing is to reliably collect and send information, e.g. by video stream or images, from source to destination. Above all, there is a need to transmit information in real time. Many different techniques and methods plagued by problems are used to process image information. One of the problems that have to be resolved is the problem of illumination variations. If this issue is not resolved, many serious problems pertaining to the reliability of the built systems, which could lead into potentially hazardous situations, could arise. For example, one can make an incorrect medical diagnosis due to the bad performance of lower vision algorithms. In order to resolve this problem, some transformation has to be used.

I. Vujović, *Multiresolution Approach to Processing Images for Different Applications*,
SpringerBriefs in Electrical and Computer Engineering,
DOI 10.1007/978-3-319-14457-3_1

This book presents an overview of wavelet approach to various problems in image processing and data visualization. A modern approach to solving the problem of illumination variations by using the time-frequency analysis method known as wavelet transform (WT) is presented. It is discussed how lower vision processing influences higher vision applications.

This book is organized as follows:

Chapter 2 deals with the concept of interaction between the efficiency of lower vision and the results at the higher level of visual applications.

Different approaches to WT realization (as an example of integral transforms) are explained in the Chap. 3. This chapter includes basic concepts and an explanation of the multiresolution idea.

The first research field presented is visual quality control. Literature overview is presented in the first section of the Chap. 4. The second part gives an experimental example of the multiresolution approach to visual quality control. Finally, the visualization of material properties is mentioned in the third section of the Chap. 4. Particularly, it states our current research topic and future development. To be more concrete, we deal with properties of the relative dielectric constant, which is important in many areas of electrical engineering, from selection of materials to communications.

Augmented and virtual realities, as parts of another research field studied by our research group are explained and some preliminaries presented in the Chap. 5. In the first section, an example of animated world is presented for security and terrorist attack detection purposes. The second section covers the issue of the health of mariners and how it could be improved by advanced communication technologies. It is difficult to predict how mariners will communicate in the future and the impact of such future communication on their mental health. There is no doubt that bleeding-edge communication technologies will have major influence on their way of life.

The most comprehensive part of our research are biomedical applications. It includes computer aided diagnostics and the role of multiresolution analysis in medical signal processing. It covers some aspects of medical imaging and EMG research. I hope I cut the presentation of this topic to reasonable measure. A multiresolution approach to EMG analysis is presented in the first section of Chap. 6. The second section deals with computer-aided diagnostics explained on an example of occupational asbestosis. The first subsection covers pulmonary X-ray compression and the influence of compression on the reliability of medical diagnostics. The second subsection deals with the visualization of asbestos infected areas intended to help medical doctors and radiologists to observe suspicious areas and determine correct diagnosis.

Multiresolution approach to different video surveillance applications is presented in Chap. 7. The first section presents experimental examples in indoor cases, and the second section in outdoor cases. The indoor examples covered in the section are robot vision and human motion detection.

Finally, the last chapter presents the conclusions.

Chapter 2
Interaction of Lower and Higher Vision Applications

Abstract In this chapter, an interaction between the lower vision application performance and the higher vision application performance is explained.

Keywords Low level vision operations · High level vision operations · Interaction · Positive feedback · Negative feedback

Figure 2.1 illustrates the interaction between lower and higher vision applications at the symbolic level. Although lower vision operations do not change the spatial size of data, they can change the color map or switch color image to gray image, which should downsize the total quantity of data.

After image acquisition, the data is entered into the image matrix. After the performance of any image processing operation (low-vision application), data remains organized in the matrix form. After lower processing, the image matrix is subjected to image analysis (high-vision application). A high-vision application yields information, which could take the form of complex data, structured data, char data or some other complex data.

There is interaction between lower and higher vision applications. Information obtained by higher vision applications can be used to narrow the area of search in the case of tracking operations. This is an example of positive feedback.

However, if a lower vision application operates badly, due to e.g. illumination variations, and produces e.g. bad segmentation, the tracker will perform worse than if segmentation was properly made. This may result in the detection of greater or smaller number of targets that there are in reality, or lead to wrong identification of targets and confusion of noise with targets.

This concept can be reduced to 1D case, for example, if electrical voltage over time is analyzed or if vibration signal of turning machine [53] is used as a diagnostic tool. In such a case, data processing results can impact reliability of the data analysis, which is higher level. In the data analysis, someone reaches the conclusions about the input data set. This conclusion can be a state, a word, a number, or anything else with smaller amount of data than the input data set. For example, we can reach the conclusions about monotonity of the function by examining time change of the function.

I. Vujović, *Multiresolution Approach to Processing Images for Different Applications*, SpringerBriefs in Electrical and Computer Engineering,
DOI 10.1007/978-3-319-14457-3_2

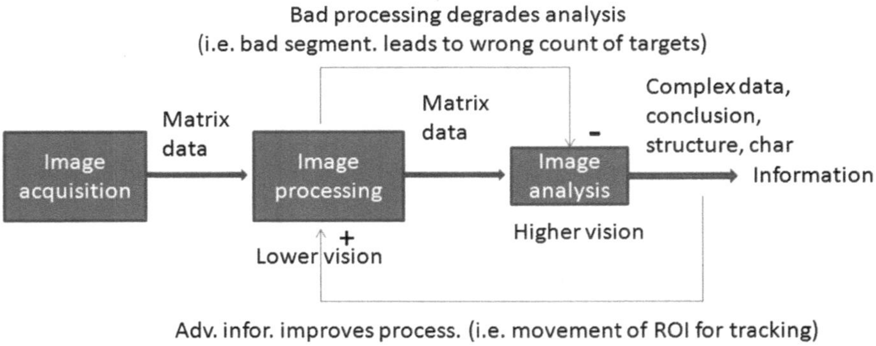

Fig. 2.1 Negative and positive influence to lower and higher vision

Generally speaking, there are a negative and a positive influence of the lower data (image) processing to the performance of the higher level applications. If the low level application (algorithm or function) operates within desired parameters, than such low level application will influence the performance of the higher level application in positive manner.

It has to be noted that higher level application can send a feedback to lower level application. The higher vision application can influence the performance of the lower level application by a positive or a negative feedback. If the higher level application operates outside desired parameters, than the negative feedback can "disorient" the lower level application. However, if the higher level application operates within the desired parameters, than the information from the higher level can improve the performance of the low level application, for example, by reducing the data set that should be taken into account by the lower level application, and, hence, increase the execution speed. This is of vital importance in online applications.

Such interaction between lower and higher level applications is important in any application, including applications based on the multiresolution approach.

Chapter 3
Multiresolution Approaches in Image Processing

Abstract This chapter explains a multiresolution approach to the image processing. Several transforms are mentioned, which can be used in the multiresolution approach. Wavelets are compared with Fourier transform, and short-term Fourier transform. Wavelet implementation issues are presented. Wavelet analysis is illustrated by an experimental example.

Keywords Wavelet · Multiresolution approach · Pyramidal implementation · Lifting implementation · Subband coding

The multiresolution approach does not only involve the repeated use of the same transform, but also the change of resolution with acceptable data loss. The multiresolution approach is usually considered to be a WT. Indeed, no one uses FT for the multiresolution approach due to the loss of the time and frequency data. However, short-term Fourier transform (STFT) can be used in the multiresolution approach to analyze non-stationary signals, such as vibration signals. In recent years, many new transformations have been proposed, e.g.:

- bandelets [54],
- complex WT [57],
- contourlets [14–17],
- curvelets [7–13],
- edgelets [1–3],
- ridgelets [55].
- shapelets [4–6],
- wedgelets [56], etc.

They are all inspired by the WT and modified to fit specific problems. All of them are intended to be used in the multiresolution approach. The common differences between wavelets and new transforms are the angle of details or filter definitions. Basically, by considering WT, we can cover all of them since some transforms are better suited for specific problems.

I. Vujović, *Multiresolution Approach to Processing Images for Different Applications*,
SpringerBriefs in Electrical and Computer Engineering,
DOI 10.1007/978-3-319-14457-3_3

There are three approaches to wavelet implementation [18–23]:

- pyramidal,
- subband coding, and
- lifting.

In the pyramidal approach, data is averaged with the neighboring data by the weighting function. The correlation of the signal with the weighting function reduces the resolution of the signal.

The point of subband coding is the division of the signal spectrum into independent subbands. Signals in subbands are treated individually for different purposes. Two-band filter bank is used for LF and HF band.

In lifting, the original signal is split in odd and even samples. The lifting scheme consists of repeated steps:

- split,
- predict, and
- update.

E.g. if Haar wavelet is implemented by lifting, the average of even and odd data has the role of approximation in subband coding. The difference of even and odd data has the role of details in subband coding.

Figure 3.1 illustrates the abovementioned concepts. Figure 3.1a illustrates the concept of the pyramidal scheme. It should be noted that downsampling process reduces the descriptive data by factor 2 in most cases. Figure 3.1b shows the concept of the subband coding. Figure 3.1c presents the lifting concept.

WT is a time-frequency technique for signal analysis. The "multiresolution" technique performs better than FT analysis for non-stationary signals, because both time and frequency information are preserved. In a way, WT is an optimized sampling. STFT oversamples the object of interest based on the Nyquist sampling theorem.

WT alters the window to overcome resolution problems. In doing so, a good temporal resolution and bad frequency resolution are obtained at HF. Good frequency

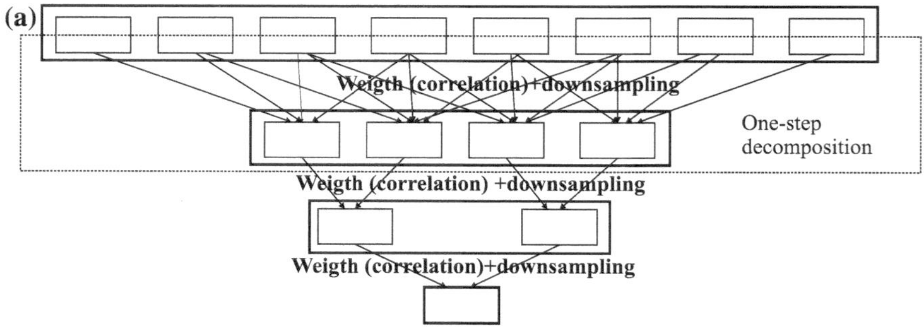

Fig. 3.1 Wavelet implementation: **a** pyramidal scheme, **b** subband coding scheme, **c** lifting scheme

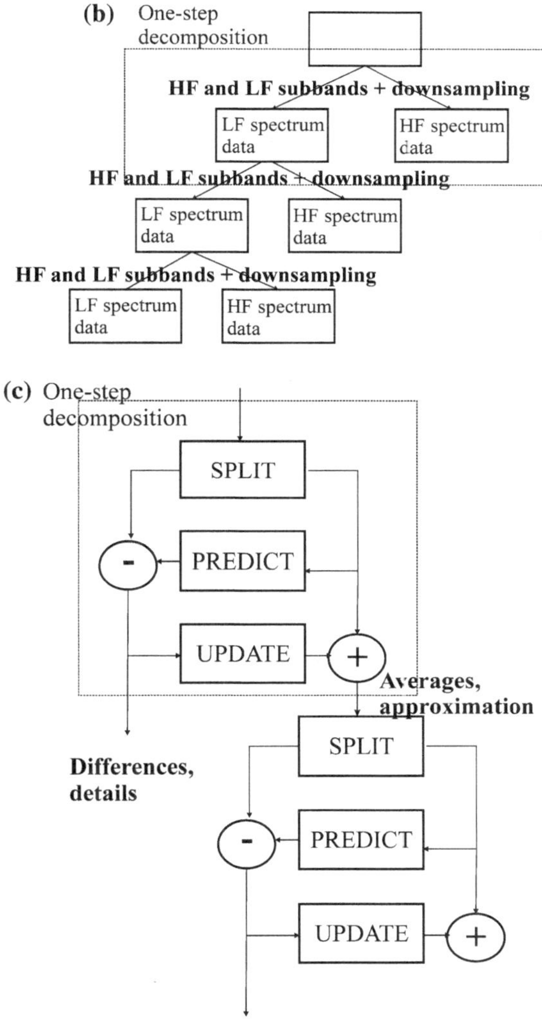

Fig. 3.1 (continued)

and bad temporal resolution are obtained at LF. However, the Heisenberg principle remains intact.

The multiresolution approach is covered in many references, including everything from mathematical approaches to different applications [24–27].

Figure 3.2 shows an example of wavelet decomposition by WT. In the first stage, the image is decomposed into:

- approximation (LL1), and
- details in horizontal (LH1) direction,

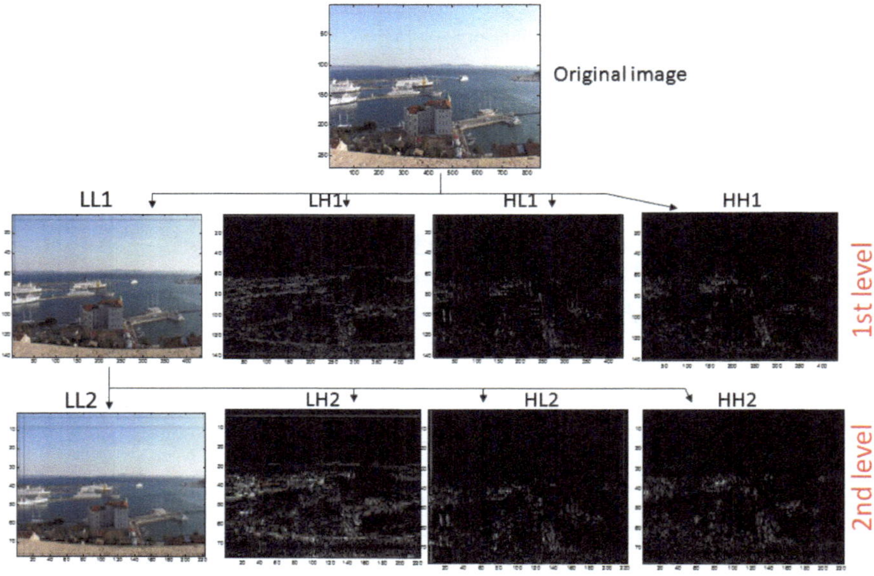

Fig. 3.2 An example of multiresolution approach—decomposition at two levels

- details in vertical (HL1) direction, and
- details in diagonal (HH1) direction.

Then, the approximation is decomposed further to obtain approximation and details at the second level of decomposition. Further decomposing of the approximation coefficients (LL1) produce split of the lower half band into four ranges that are:

- approximation of the approximation (LL1) from the first level (LL2),
- details in horizontal (LH2) direction from the fist level approximation (LL1),
- details in vertical (HL1) direction obtained by the decomposition of the approximation at the fist level (LL1), and
- details in diagonal (HH1) direction obtained by the decomposition of the approximation at the fist level (LL1).

At every level, the amount of data to be analyzed is reduced. It can be seen that human eyes, at first sight, cannot determine the difference between the original image, LL1 (which has a half of the rows and half of the columns of the original image) and LL2 (which has a half of the rows and half of the columns of the approximation image, LL1). See Fig. 3.2 once again.

Figure 3.3 shows how even the superresolution problem can be seen as an inverse transform problem, for example inverse WT (IWT). In this case, details (details coefficient matrix, to be exact) are obtained by a noise model and the simplest way is to set details to zero, which means that the assumption is that there is no noise. Hence, the "superresolved" image appears darker. This is not a real superresolution, but it can be called a quasi-superresolution, hence there are no

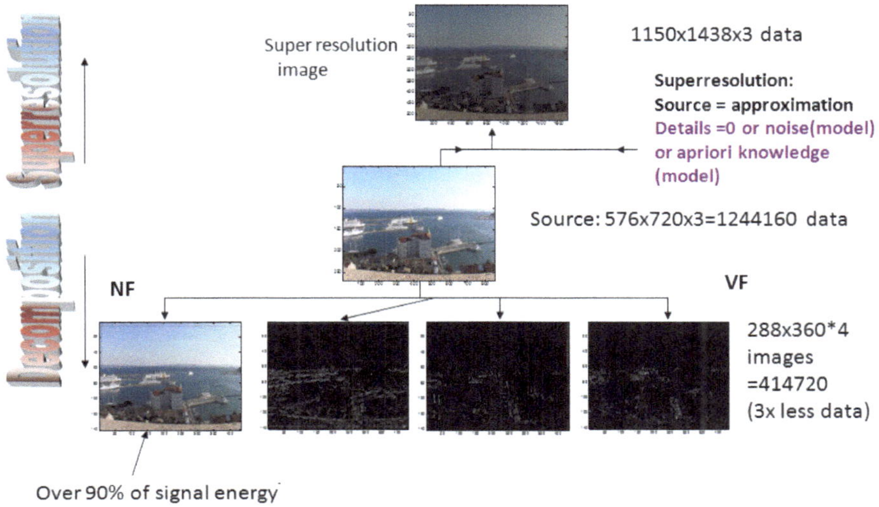

Fig. 3.3 Similarity of problem of superresolution by WT and IDWT

two or more low-resolution images which could be used as source for the high resolution image. Quasi superresolution is defined in several research papers of our team [69, 70].

Additionally, the source image is used as the approximation (approximation coefficients matrix, to be exact). Brightness and sharpness of the superresolved image depend on noise/details model/prediction.

Problem of decomposition by an integral transformation or a problem of a synthesis is in the multiresolution approach the same, just moving in opposite direction.

Multiresolution approach owns its popularity to such flexibility. Additional point in the flexibility is a fact that over 90 % of all signal's energy is contained in the approximation coefficients. Hence, in many applications is possible to reduce amount of data, not just by downsampling, but also by even totally neglecting the details coefficients (details are usually noise, although, in some cases, can contain a useful information).

The multiresolution approach can be used in 1D signals. At the low resolution, the entire signal is examined. At the high resolution, the chosen parts of interest are examined in details. For example, one can detect small surges, which should be analyzed and increase the resolution in small time intervals when the surges occur.

This book mostly deals with the multiresolution approach in case of images. This approach will be presented at the examples from researches of our team in following chapters.

Chapter 4
Visual Quality Control

Abstract Quality control is an interesting topic in signal processing. This chapter deals with quality control with a camera input to the algorithm. A literature overview is given. It is presented how the multiresolution can be used in this application. The last section explains how to use visualization techniques in dielectric material's properties analysis.

Keywords Energy measure · Normalized wavelet approximations · Noise · Material's properties visualization · Relative dielectric constant

In this chapter, I will present a literature overview of the visual quality control systems, with the accent to the multiresolution approach. It is presented that there are not a lot of researches with the particular application of the multiresolution approach or, specifically, wavelet applications.

The second section presents the research of my research team in this field. Finally, some directions for further research are presented in the last section.

The constant of this chapter is in the research of material's influence to the final results. The final results are obtained by the multiresolution approach. Wavelets were used in the presented cases.

4.1 Literature Overview

An interesting application of computer vision is the development of automated visual control quality systems. There are references about this topic. For example, visual inspection of wire bonding is described in [28]. The authors wanted to improve computer vision technique to be at least as good as human perception. Visual inspection of metal parts was reported in [29]. This system recognized several possible defects through an image analysis algorithm. A vision system for plank tracing in sawmills is presented in [30]. The system uses cross-correlation in phase in the FT domain. Matlab was the test software. Visual quality inspection

I. Vujović, *Multiresolution Approach to Processing Images for Different Applications*, SpringerBriefs in Electrical and Computer Engineering,
DOI 10.1007/978-3-319-14457-3_4

in cast iron production is elaborated in [31]. Machine learning methods were implemented and compared. An intelligent vision system for ceramic tile industry was proposed in [32]. The intelligent part of the system is included by neural network classifier (NNC). The surrogate feature vector is introduced in [33], in which wavelet multiresolution properties were successfully combined with discriminative properties of Zernike moments. However, this approach also uses NNC. It was reported that 192 instead of 256 moments were sufficient and more efficient for different robot vision and visual quality control applications. However, only the Haar wavelet was used. Structural damage health monitoring performed by image processing is presented in [34]. Displacement of control points was estimated by the optical flow method. Another civil engineering problem solved by image processing is facilitation with the estimation of damage caused by natural disasters [35]. It was performed by detecting changes on satellite images.

The Empirical Mode Decomposition (EMD)-based Wavelet damage detection method was developed for damage detection (both time and location of damage) [36]. In the same dissertation, the wavelet-energy (WE) based system identification method was developed. WE is used to identify physical parameters of the structure. Surface failures on ceramic tiles were investigated in [37].

DWT was used in preprocessing and ANN for the detection of surface defects. A defect detection computer vision system is presented in [38]. It quantifies different types of ceramic tile defects. It was tested in industry. An integrated quality control industrial system is presented in [39].

3D was obtained by monochrome single camera and 6 DOF robot mounted laser scanner probe device. The approach is not, however, mulitresolutional. Surface roughness metrology is presented in [40]. It is based on high quality images and complex wavelets used for machine components, improving corrosion resistance, creep life, fatigue strength and other factors.

Structural health monitoring (SHM) in civil engineering can be performed by a visual system, as in [41]. Displacement fields were calculated by the so called image correlation coefficient.

This literature overview shows the diversity of applications of computer vision-based quality control systems. The next section shows an example of the multiresolution approach to a visual quality control. This example is based on the wavelet multiresolution approach.

4.2 Example of Multiresolution Approach to Visual Quality Control

There are several points at which the multiresolution approach can be introduced into visual quality control.

The first is the use of the transformed domain to reduce the amount of data to be analyzed by other means.

The second frequent use of the visual quality algorithm is in data (images) preprocessing, e.g. denoising (prefiltering).

The third is the use of the transformed domain for feature extraction, leading to image analysis (higher vision application).

The natural extension is the combination of a low-vision application (filtering) with a high vision application (feature extraction which leads to the final conclusion). Although the use of the transformed domain is mystified to an extent, it is basically performed to facilitate the task.

Figure 4.1 illustrates the influence of external conditions on image quality. Figure 4.1a shows the image of a product. Figure 4.1b shows the same product under non-uniform illumination variations. Figure 4.1c is an attempt to visualize only illuminations, which can be considered as a type of noise. The results of visual quality control can be assumed to vary depending on the level of the illumination variations.

Wavelets were observed to increase the tolerance of visual quality control systems to error. As experimentally verified in [42], in the original image, tolerance to error is only 1 %. As shown, the used wavelets are best suited for speckle noise.

The results [42] mean that the image of a product can be changed by as many as 25 % (in case of speckle noise) without influencing the correctness of product classification (for possible removal).

In order to evaluate and obtain the above mentioned results, energy measure (*EM*) is introduced in [43]:

$$EM = \frac{1}{N}\sum_{i}\sum_{j}\sum_{k} a(i,j,k)^2 \tag{4.1}$$

where:

- *EM* designates the energy measure,
- *N* the number of pixels in the ROI for the considered level of decomposition,
- *i* and *j* spatial coordinates of the pixel for which *a* is energy, and
- k the color designation (red 1, green 2, and blue 3).

Therefore, $a(i, j, k)$ is the energy of the pixel contained in the *k*th color calculated in Hilbert space (wavelet domain). Energy measure is used as a number for comparison of energies of damaged and undamaged ROIs. EM is calculated in Hilbert space using Parseval relation [43]. The calculated energy is divided by the number of coefficients in the ROI.

Possible computer vision algorithm problems are: edge detection, color conversation for edge detection purposes, ROI shape modeling, ROI recognition, ROI analysis with conclusions about quality, image measurements, etc. Since, this is not a self-learning method, it greatly depends on the reference model. If there is a new product, an anomaly cannot be detected if the desired image is not already in the memory.

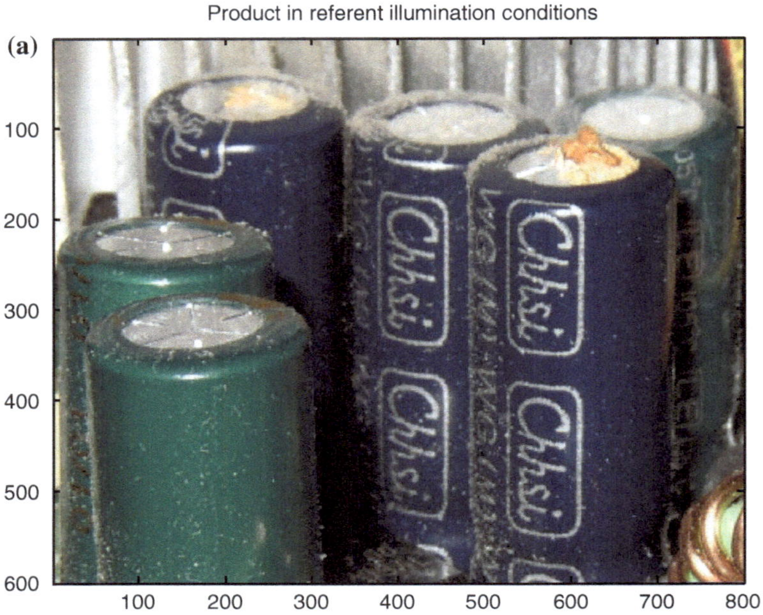

Fig. 4.1 Example of electronic circuit: **a** clear image, **b** image under non-uniform illumination variations, **c** visualized variations

It can be shown that illumination variations are a zero-sum game if a sufficient number of frames are used in the analysis. The idea is to use several frames and suppress the influence of illumination variations.

In order to experimentally illustrate expectations, the following experiment was executed. The original image (Fig. 4.2a) is artificially changed. Global illumination variation is simulated by adding a constant brightness of 40, -100, and 60 to the original image. Intuitively, the sum of changes is zero. However, the problem is that the product must not be moved through the scene prior to the evaluation completion. But since an automated system only requires a few frames, a small portion of a second, this should not be a serious problem.

Figure 4.2b shows the average of variation-poisoned images. There are some differences from the original. Figure 4.2c shows the absolute difference of Fig. 4.2a, b. Figure 4.2d is a histogram of image differences (Fig. 4.2c).

The difference can be seen to be in the lower part of the spectrum. Therefore, the LF filter can logically be expected to suppress the differences, which should virtually be set to zero. The problem with such intuitive evaluation is in the fact that images have limited possible pixel values. The values should therefore be normalized and the loss effect should be included in the calculations.

Figure 4.3 is an example of what happens when an image is subject to illumination variations. Figure 4.3b shows wavelet approximation coefficients of the image under illumination variations. Figure 4.3c shows the difference between

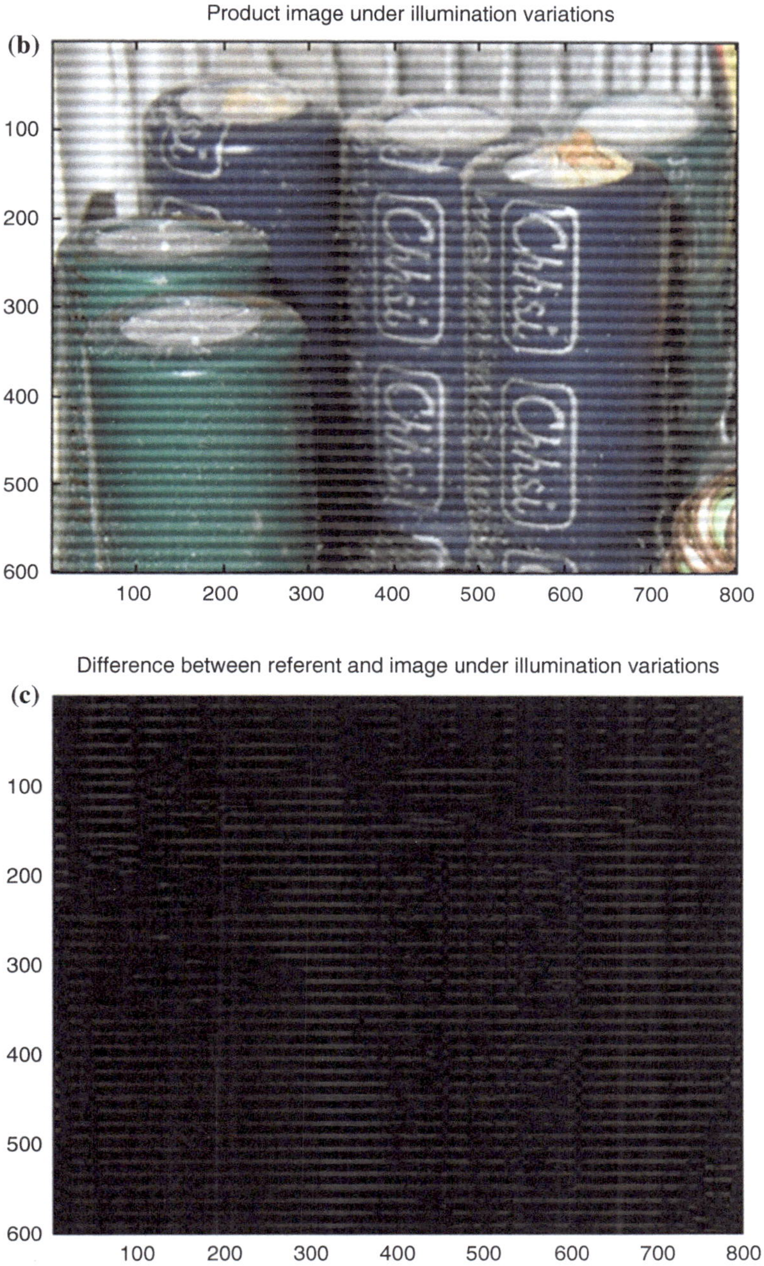

Fig. 4.1 (continued)

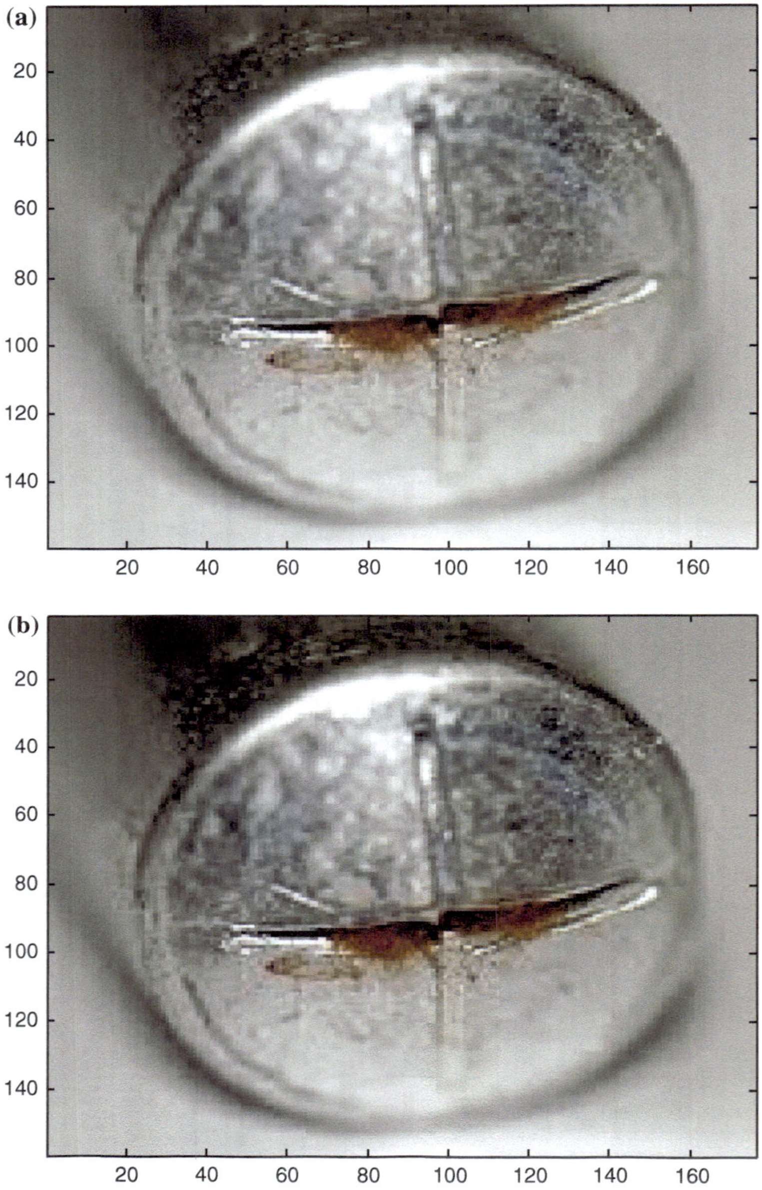

Fig. 4.2 **a** Original image, **b** average of images with three added variations, **c** difference between the original and the average of images with three added variations, **d** histogram of differences

(c)

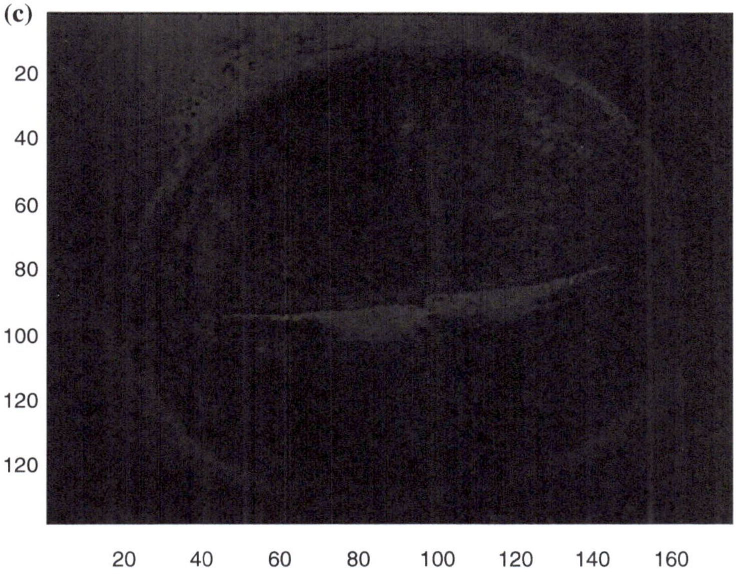

(d)

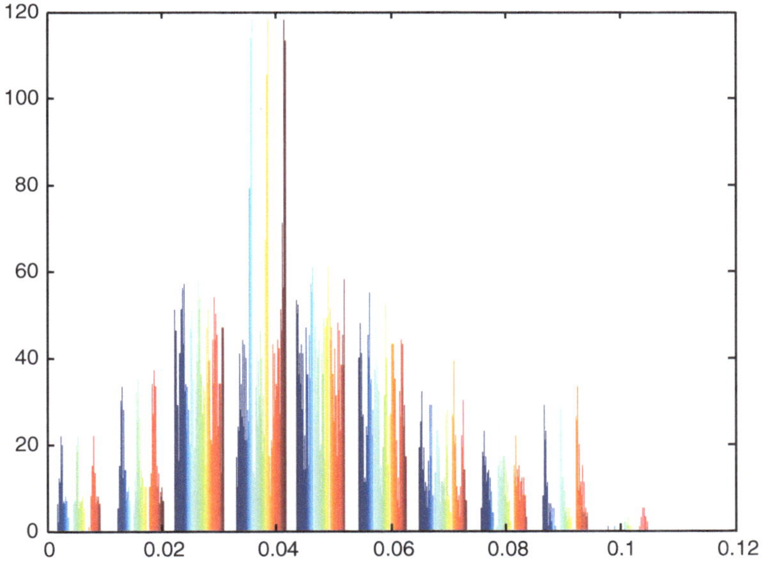

Fig. 4.2 (continued)

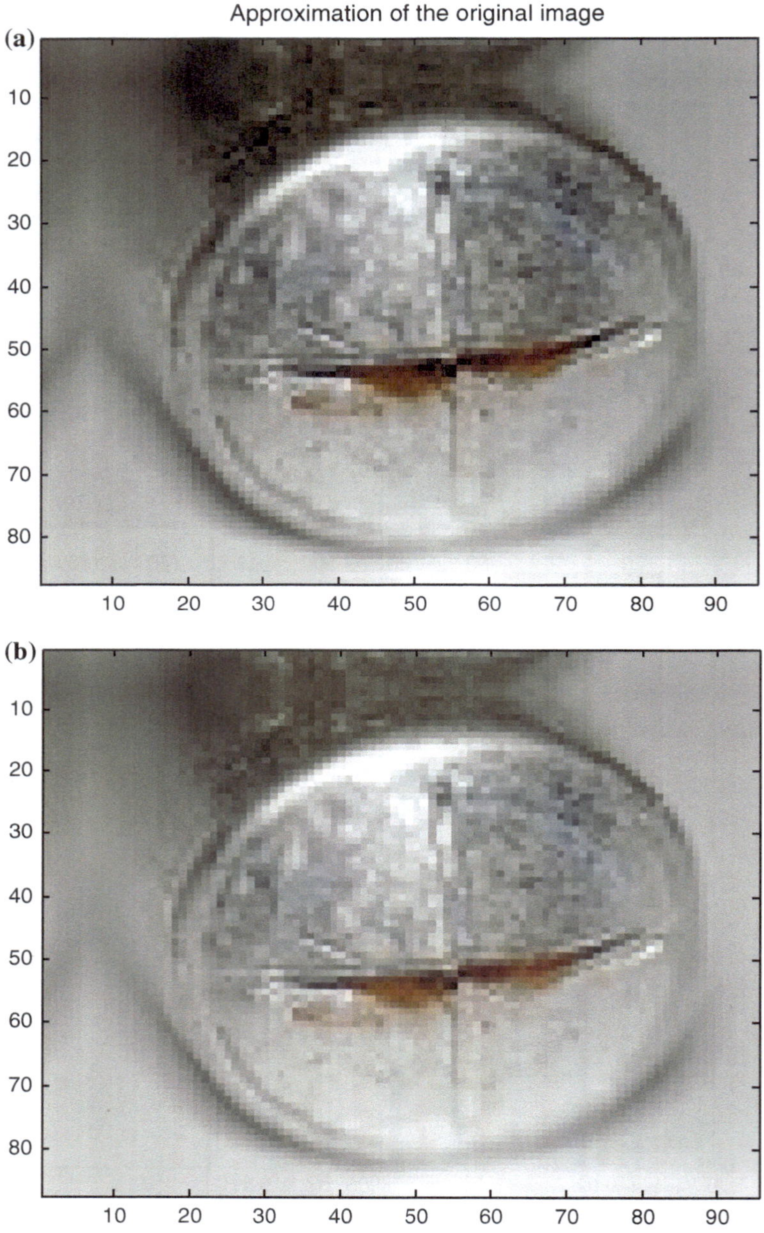

Fig. 4.3 **a** Wavelet approximation of the original image, **b** approximation of the illumination-added image, **c** difference between (**a**) and (**b**), **d** histogram of (**c**)

(c)

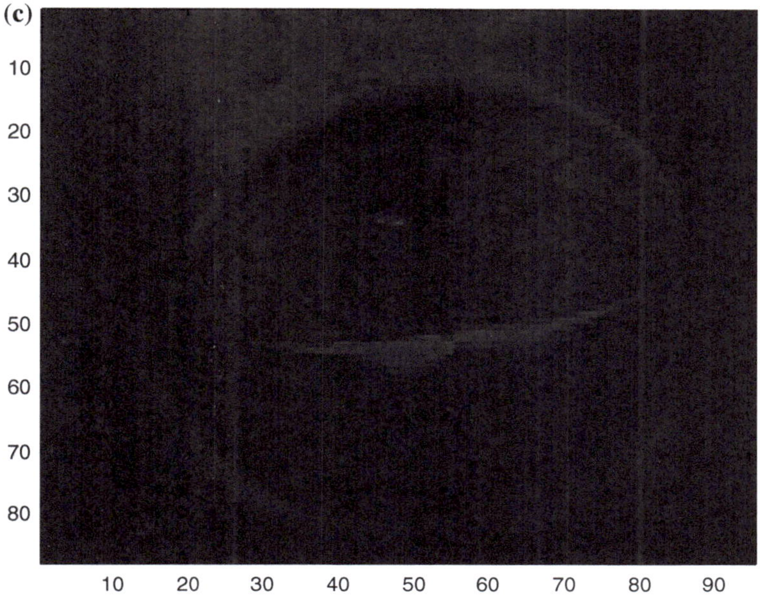

(d)

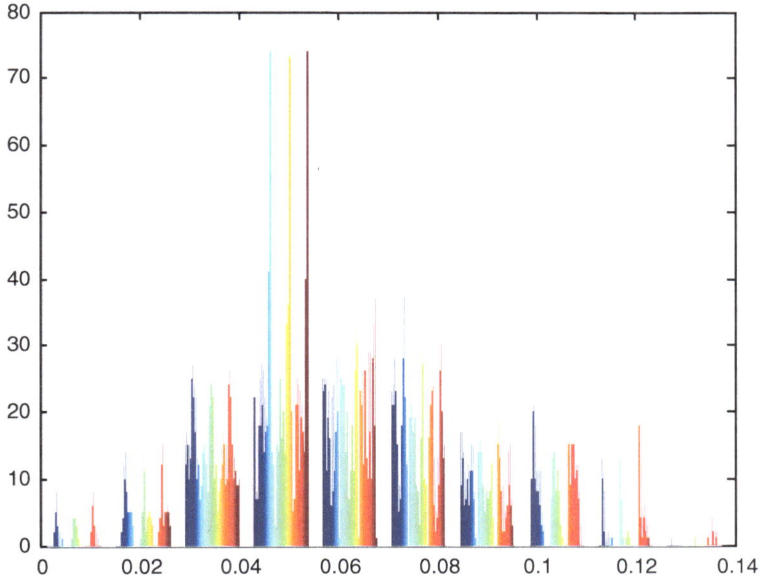

Fig. 4.3 (continued)

the approximation of the original and illumination-added image coefficients. Figure 4.3d shows the histogram of differences.

The wavelet filter (in this case dB8) can be seen to produce wider spectra (bandwidth) of differences and lower amplitude. Of course, this is an image with only one added variation.

The next stage in our experiment was to include all images with three added variations into the simulation. In order to obtain an average, we should perform the following operation:

$$r_1 = \frac{|a_n + b_n + c_n|}{\max(|a_n + b_n + c_n|)} \tag{4.2}$$

where a_n, b_n, and c_n are normalized wavelet approximations of illumination-added images.

Figure 4.4c shows that the result of the described procedure reduces amplitude and bandwidth by examining the histogram of the normalized sum. Figure 4.4a, b shows addition of illumination variations and the image difference.

Table 4.1 shows results obtained by our previous researches in this field. It is tested how the performance of the visual quality control system is influenced by three typical types of noise. Algorithm is based on the multiresolution algorithm developed in [58]. It is, essentially, an energy-based wavelet algorithm.

Normalized average region of interest (ROI) energy is calculated for a satisfactory product (made within fabrication tolerance) and for an unsatisfactory product. The calculated energies are presented in the Table 4.1. It can be seen that the highest tolerance to a noise is to, so called, the speckle noise. This was unexpected conclusion at the time of the research. It can be noticed that tolerances to the Gaussian and to the Salt and Pepper noise are relatively small. This could indicate that wavelet energy algorithm is not satisfactory for every types of noise. So, the algorithm parameters should be changed, such as different thresholds, levels of decomposition, wavelet family or number of moments.

The presented method has disadvantages at this stage of development. Firstly, numbers for normalized ROI energy vary from case to case. It is possible to conclude what is a good product and what not, but not based on uniform measure, which should be deduced. This is a problem, because there is no referent range of numbers for comparison. Hence tolerance to noise is low, it will be hard to realistically determine range of energies which are for good and for bad products. So, it is imperative to generate a reliable energy measure for this application.

The second problem is in investigation of different operating conditions. It is not possible to expect the same condition all the time. Stochastic disturbances can make an algorithm to fail. These disturbances should be minimized to increase the reliability.

Low-level processing must be implemented to reduce the influence of illumination variations (which are, in fact, one type of the stochastic disturbances) on the performance of visual quality control systems. A similar conclusion can be reached in the case of the influence of different types of noise or any other low level phenomenon. Consequently, bad low-level processing performance can only

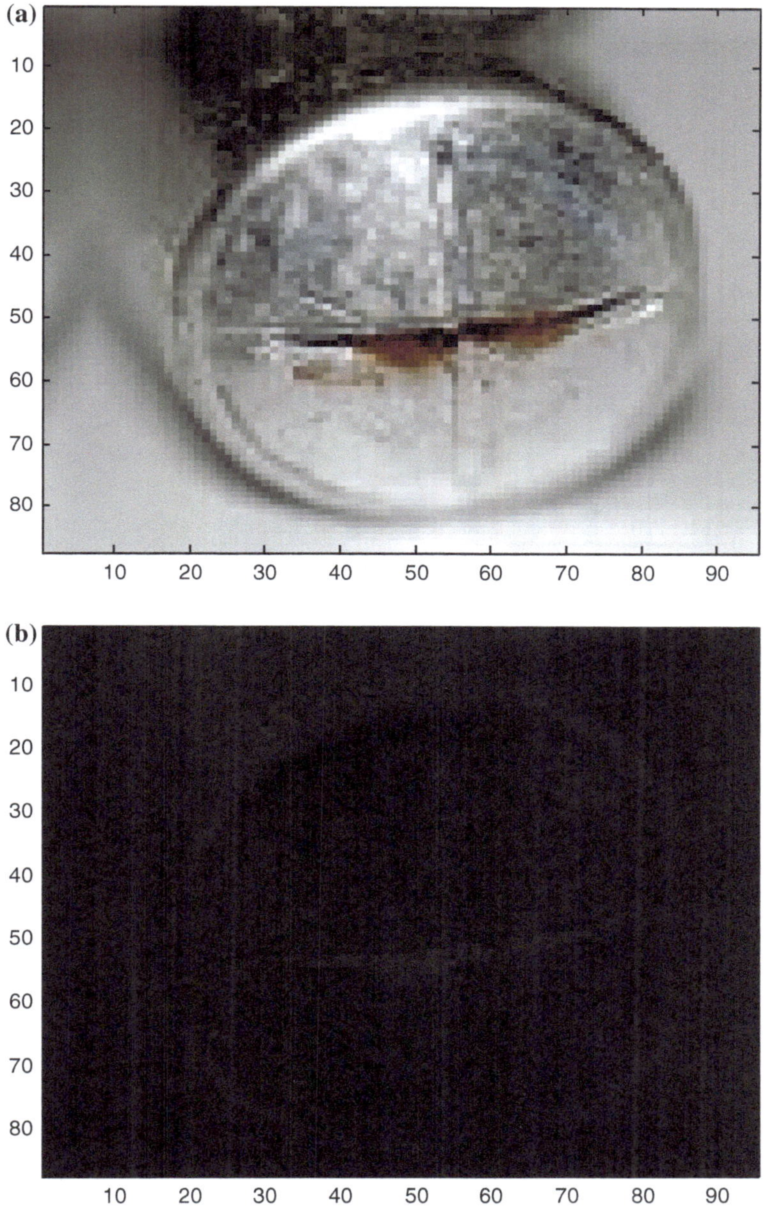

Fig. 4.4 **a** Normalized sum of variation-added approximations, **b** the difference between the normalized sum and approximation of the original image, **c** histogram of the normalized sum

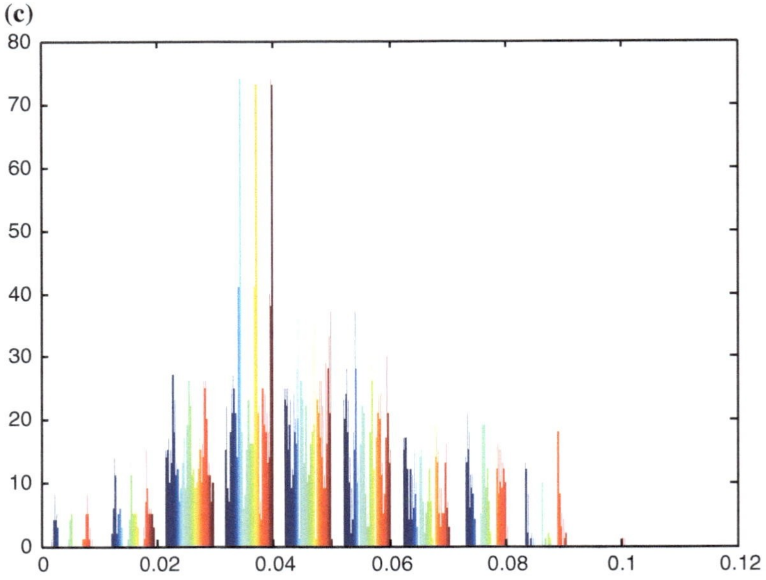

Fig. 4.4 (continued)

Table 4.1 Numerical experiment results for robustness of the visual quality control system to noise

Noise type	Speckle noise	Gaussian noise	Salt and Pepper noise
Normalized average ROI energy for satisfactory product	0.6743	0.6115	0.5517
Normalized average ROI energy for unsatisfactory product	0.5046	0.5780	0.5374
Tolerance to noise (%)	25.167	5.478	2.592

lead to poorer high-level vision application performance. In this case, illumination variations can be concluded to influence product quality evaluation.

The further research should be in investigation of possibility to find a reliable energy measure and to implement the finalized algorithm in industry.

4.3 Visualization of Dielectric Constant Dependences

Data processing can be used to visualize material's properties. The result can be in making conclusions easy to comprehend. So, the low level processing of data can result in high level conclusion about material's properties. Examples are given in [60, 61]. Material's properties are important in the quality control, because it can be possible to predict the product's quality from the material's properties.

For example, some technological procedures can change material's properties and lead to defects. If these defects are not taken into account, then the product can fail to perform due to the changed dielectric constant or the dielectric breakdown.

Typical influential parameter is the moisture. The relative dielectric constant is dependent of the moisture. The moisture can be found in the soils, the solid materials, gas dielectrics, etc. Absorbed moisture during production can make fatal defects in materials crystalline structure. Such defects usually increase chances for the dielectric breakdown. If it is possible to detect such fatal defects in the process of the quality control, it would make products that come out of factory more reliable.

Changes in the value of the complex relative dielectric constant are not solved in general by theory. It depends on materials type. There are studies that provided the results for different types of i.e. soils.

According to [62], the real part of the dielectric constant increases exponentially with moisture percentage. Tangent of dielectric loss or imaginary part of dielectric constant increases linearly with volumetric moisture content.

Another interesting research [63] shows that electromagnetic transparency increases by drying the material and that transmitted power decreases if moisture increases, which is linked to the dielectric constant. It was concluded that in X-band (a high frequency band used in satellite communication) exists direct relation between moisture and dielectric constant, expressed with [63]:

$$\mathrm{Real}(\varepsilon_r) = 3.95\exp(2.79Mc) - 2.25 \tag{4.3}$$

$$\mathrm{Imag}(\varepsilon_r) = 2.69\exp(2.15Mc) - 2.68 \tag{4.4}$$

where numbers depends on material and these numbers were obtained for the researched material [63] by experiments. Mc is the moisture content.

According to [64], porous low-value dielectric constant materials absorb more moisture. The chemically absorbed moisture degrades electrical and reliability performance of the dielectrics with low value of the dielectric constant.

A model of volumetric moisture content influence to the dielectric constant is evaluated and confirmed in [65]. It is polynomial dependence of the first or the third order, expressed with:

$$\theta = a \pm b \cdot \varepsilon_r \tag{4.5}$$

$$\theta = a - b \cdot \varepsilon_r + c \cdot \varepsilon_r^2 - d \cdot \varepsilon_r^3 \tag{4.6}$$

where *a*, *b*, *c* and *d* are dependent on the material's type. These empirical equations were confirmed in some types of soils, but not in electrical products with great influence of the relative dielectric constant.

Since, my team is stationed at the Maritime Faculty, we are especially interested in the influence of the abovementioned to the marine applications. It cannot be emphasized enough how important such research are in maritime affairs. Ships and offshore facilities are under a high impact of the moisture due to high humidity in and around sea water.

One of electrical engineering fields is a field of satellite communications and Internet, radar satellites, GPS, etc. This field is increasingly interesting in marine applications due to more advanced application, which are wider every day. In [66], an influence of moisture content to the ionosphere disturbances is considered for the X-band. It was concluded that such an influence can be expressed with:

$$\varepsilon_r = \left(a \cdot e^{b \cdot Mc} - c\right) + j\left(d \cdot e^{e \cdot Mc} - f\right) \tag{4.7}$$

where a, b, c, d, e, and f depend on the material of the communication part considered, and Mc is the moisture content.

For example, if some communication equipment is made from wood (depending on wood's type), these constants can be, for example:

- $a = 3.95$,
- $b = 2.79$,
- $c = 2.25$,
- $d = 2.69$,
- $e = 2.15$, and
- $f = 2.68$.

Multiresolution can be used in:

- curve/surface fitting of different dependencies,
- visualization of a part of the range, which is of interest in specific application (the whole experimental data can be in shown in one resolution, and the specific range of data in the different resolution),
- estimation of different parameters,
- solving the equations of dependencies, etc.

This is the research started in the last year (2013). There is no way of telling would it be successful. However, this chapter let us aware of materials' influence to the final properties of products and to the quality of the manufacturing processes.

Chapter 5
Possibilities for the Improvement of Human Work Conditions by Virtual/Augmented Reality

Abstract This chapter deals with possibilities for the improvement of human work conditions and with the relaxation at remote and isolated work places (oil rigs, ships, spacecrafts, etc.). It is shown how cutting-edge visualization and communication techniques can improve safety and security. Furthermore, a multiresolution role is explained.

Keywords Virtual reality · Animated world · Mariners' health · Communication systems · Security

Another example of our research is advanced visual techniques. There are references about the use of augmented/virtual reality in medicine, telesurgery, or robot control [44–47]. These applications are well-understood. They are mostly based on satisfactory communication systems. If communication fails, the application fails. Since in some applications, such as treatment of stroke patients, the application can be performed in situ, communication between different parts of the world plays no vital role.

Possible applications of virtual reality range from serious applications to entertainment industry. Examples of serious applications are:

- search and rescue,
- distant robot control, or
- flight simulators, and so on.

Medical applications are telesurgery or rehabilitation. In the entertainment industry, applications are in:

- games,
- holonovels,
- interactive novels, and
- future movies.

The development of such systems is sponsored by the largest industries, such as military, biomedical and entertainment.

I. Vujović, *Multiresolution Approach to Processing Images for Different Applications*,
SpringerBriefs in Electrical and Computer Engineering,
DOI 10.1007/978-3-319-14457-3_5

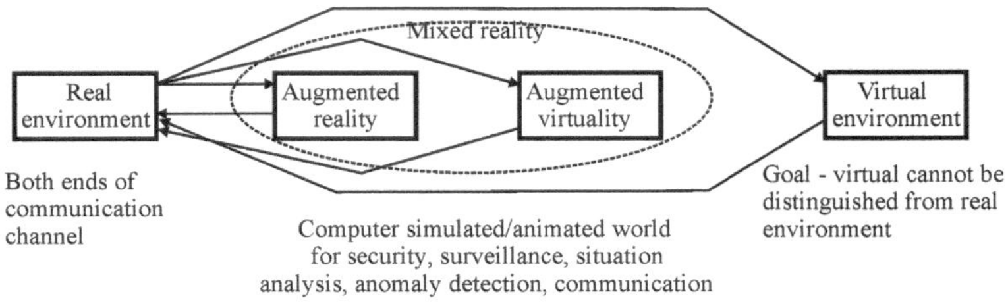

Fig. 5.1 From computer animated world towards virtual reality

Basically, advanced visualization techniques are categorized as augmented reality, augmented virtuality or virtual environment [46]. Figure 5.1 shows the interaction of real and advanced visualized worlds. It should be noted that due to the advancement of other sensors, such environments cannot rightfully simply be called visualized worlds. It is interesting to point out the advancement in tactile sensors. Other integrated sensors are audio and recently, smell. Some researches about flavor have also been made in the food industry. The integration of all sensors will make virtual reality possible which could be considered a substitution for the real senses and the real world. Furthermore, it would allow the uploading of a person's mind into a virtual reality computer system and make possible immortality sort of possible, which could be e.g. therapeutic in case of terminal illnesses.

Any form of advanced visualized world is based on communication. Both ends of the communication channel end in the real world. In the beginning, there is an actual stimulus in the real world. At the end, an actual reaction is produced that affects the real world. In order to visualize the situation, a human operator is under visual (and other senses in the future) stimulation, which is augmented to another space to facilitate the task/decision. There are two steps of augmentation:

- augmented reality, and
- augmented virtuality.

They are both called mixed reality. In mixed reality, the operator is aware of both the real and the augmented world, but gets the results by concentrating. In virtual reality, it should be impossible to distinguish the real world from the virtual world, because all senses are involved in the virtual environment and the operator's real world location is isolated from real-world stimuli. However, this high level of virtuality is not necessary to perform the desired tasks.

A possible application of virtual/augmented reality is to make glasses with small cameras, which would create the illusion of 3D space. A small processor can take care of edge detection and recoloring of full planes. This output should be connected to the nerves of people suffering from visual disabilities to make their lives easier. That is an example of biomedical application. In this case lower vision (edge detection, plain close, recoloring) should be used to enable the human brain to interpret them at higher vision, e.g. to avoid obstacles or find a way home.

Once again, it should be aware of the interaction between lower data processing and higher level applications. In this case, sensors are usually visual in nature, such as a CCD. Low level image processing is performed between visual sensors and an input to the higher vision application, such as a computer code for animation. If a lower vision performed its task unsatisfactory, the result will be confusion for a user of such an animated/augmented world. For example, colors can be changed rapidly. This will make harder to an operator to focus to the essence of the application. As a feedback, a command sent back to a virtual/augmented/animated world control application could be seen as unsatisfactory reaction.

Multiresolution is included through the data processing part between the sensors and the control application. In another part of the system, multiresolution application can be in making higher resolution in focused part of the virtual/augmented/animated world and lower resolution to a part of the world marked as unimportant background. It should be noticed, that the background could be classified as the important and as the unimportant. The important background could interact with objects in focus. It is, essentially, a way human brain operates: we do not notice usual, irrelevant details. However, major thing are noticed in the best possible resolution.

Two examples are presented in this chapter:

- example of animated world in security application, and
- analysis of the influence of the improved communications to mariner's health.

5.1 Animated World Example

Primitive virtual/augmented world is the animated world. Functions of virtual, augmented and animated world could be the same. However, operator can get into the spirit in the virtual world easily and in the animated hardly, because animation is considered as unnatural.

The following example is a case of the application in port's security. Animated world discards everything irrelevant letting the operator to focus on the real treat.

Figure 5.2 illustrates anomaly detection for security applications in the animated world, which is a low-level augmented reality [48]. Potential terrorist is detected by differencing the referent model and the visualization of the real scene currently obtained by sensors.

Unimportant (irrelevant) parts of the background are:

- most of the sea surface, and
- most of the dock surface.

Parts of the background that should be in focus are:

- dock's part close to the ship and ship's trajectory,
- proximity of the ship and the tug boat (to avoid collision), and
- possible anomaly (or anomalies) in case of Fig. 5.2b.

Operator's role is to distinguish between treat and non-dangerous anomaly, such as tourists, fish boats, boats in stress, or else.

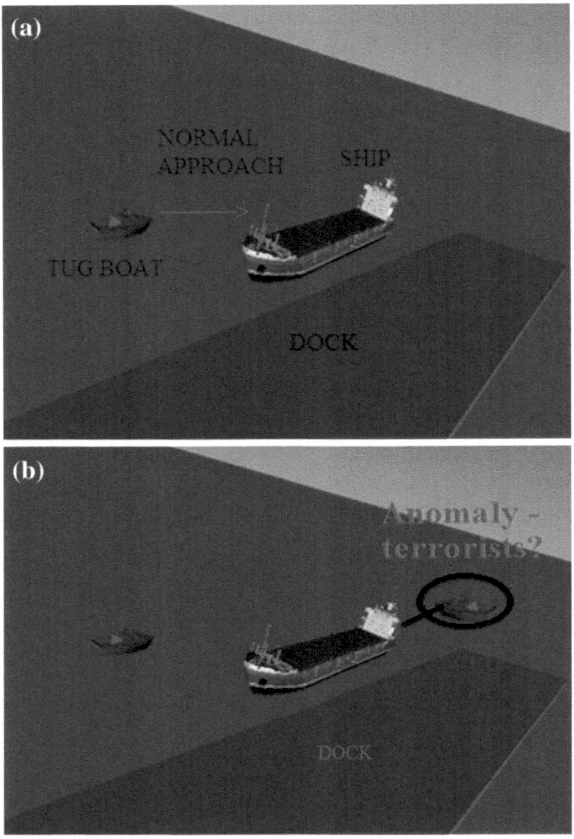

Fig. 5.2 Potential terrorist attack detected in animated world: **a** a normal approach to the arriving ship, **b** unusual situation—anomaly, which could be terrorist attack

Interaction between lower and higher vision application is obvious. In order to operate faster, it is necessary to reduce data to be processed. It is performed by extending the area of the lower resolution (irrelevant parts of the background). However, reduced resolution can cause inability to detect small changes (i.e. a small boat), which could be potential treats. Further research of our team in this research field is to add a learning quality to vision algorithms, which will incorporate an optimal approach between the lower resolution parts and the anomaly detection.

5.2 Improved Communications in Maritime Affairs

In this section, an example of advanced visual communication is considered. We will discuss how such techniques can be used to help seafarers suffering from the psychological trauma of separation from home [49].

The example of human error [49], which may be caused by fatigue and excessive workload and lead to captain's perceptual error, is collision of the M/V Santa Cruz II and the USCGC Cuyahoga on a clear, calm night on the Chesapeake Bay. The most fatal result was lost of lives of 11 Coast Guardsmen. It is difficult to take account all the effects which could influence seamen psychology. One typical influence is the noise.

The noise is caused by different sources, such as ship's engines, generators, air-conditioning, etc. It can cause problems even if cannot be heard, i.e. in [49]:

- thorax (3–7 Hz),
- heart (4–8 Hz),
- abdominal and thoracic organs (4–9 Hz),
- spine (2–6 Hz), and/or
- pelvis (4–9 Hz).

Head problems are caused by frequencies between 20 and 30 Hz.

Due to the advancement of modern communication systems, seafarers are no longer under so much pressure, because of, for example, telemedical care, including consultations, counseling or telesurgery, and improved physical trauma treatment aboard ships.

Psychological traumas have also been observed [50, 51]. The anticipated benefit of improved communications is the easement of psychological problems caused by separation. The extent to which modern communication improved the life of seafarers over the last century cannot be stressed enough. Even in the twentieth century (except in the last decades), personal communications were limited and a cause of stress to seafarers. Nowadays, seafarers have cell phones at their disposal with satellite links, which enable communication at almost any time. Furthermore, modern ships also have available Internet. It is not only possible to use e-mail communication, but also to see each other through applications such as Skype.

Future trends in communication technology include [49]:

- 3D video conferencing as an improvement of the present day videoconferencing,
- 3D video phone,
- virtual reality with haptic interface,
- holographic visualization in 3D space, etc.

New technologies should be approached with caution. A Samsung report [52] states that possible symptoms of watching 3D pictures include:

- altered vision,
- lightheadedness,
- dizziness,
- involuntary movements such as eye or muscle twitching,
- confusion,
- nausea,
- loss of awareness,
- convulsions,
- cramps and/or
- disorientation.

Seafarers are not a risk group, rather the problem lies at the other end of the communication channel. The identified risk groups are [49, 52]:

- pregnant women,
- young children,
- teenagers,
- the elderly,
- people prone to seizures or stroke,
- people prone to dizziness or motion sickness,
- people with eye problems,
- people who are out of shape, and
- people who have been drinking.

This opens the question of whether new technologies are safe for communication between a pregnant wife and husband on an off-shore job.

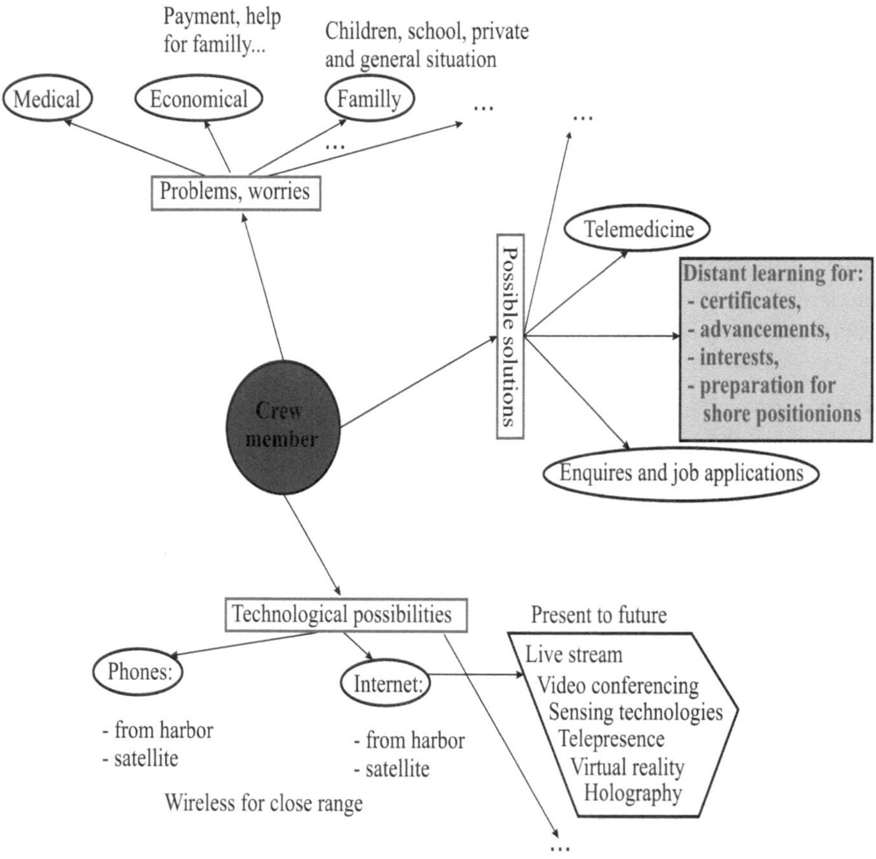

Fig. 5.3 Possibilities of new visualization technologies for seafarers

Figure 5.3 shows the possibilities of new visualization technologies in the improvement of seafarers' lives (see [49]). The currently available technological possibilities are phones and the Internet. Phones and the Internet can be used from the harbor or by satellite. Future technologies are:

- live stream over satellite from any part of the Earth,
- video conferencing (3D),
- sensing,
- telepresence,
- virtual/augmented reality, and
- holography.

Possible applications are in medicine, economy and family lives. Modern technologies will enable distant learning. This can be useful for seafarers, as it would allow them to obtain certificates and prepare for shore jobs. Seafarers could make inquires and submit job applications when not on duty. Figure 5.3 illustrates just how much can be achieved with advances in this field, which could reduce and virtually close the gap between off-shore and on the shore jobs.

The multiresolution approach can be used in several ways:

- coding of data to preserve confidentiality and privacy of communication and/or transmitted data (wavelet coding),
- compression of data to increase transmission speed (it is especially important in high resolution visual data, which are large pockets of data and in a teleoperation applications),
- filling of missing data pockets,
- prefiltering,
- postfiltering, etc.

The interaction between lower and higher vision is obvious in case of data compression for transmission from land to ship and vice versa. It should take greater role in future communications. Such an interaction can also increase efficiency of the algorithm for filling of missing data. It is a way for further researches in this research field of our team.

Chapter 6
Computer-Aided Diagnostics and Multiresolution Analysis Role in Medical Signal Processing

Abstract Multiresolution role is considered for the application in computer-aided diagnostics. Two examples are shown, which are in the scope of our research group. The first example presents preliminary results in EMG analysis of brain signals obtained by experimenting with stutterers. The second example shows research results in diagnostics of occupational asbestosis. It was found how much levels of wavelet decomposition can be used without damage to the medical information in pulmonary X-ray images of the asbestos infected patients.

Keywords Computer-aided diagnostics · EMG · X-ray images · Image compression · Stutterer · Asbestos

An interesting research field deals with biomedical applications. Biomedical applications could be:

- analysis of medical images,
- help in diagnostics by emphasizing certain data or by suggesting possible actions or diagnosis,
- archiving biomedical data in compressed digital form, etc.

In this chapter, we will discuss two examples:

- preliminaries in electromyographic (EMG) of stutterer's brains, and
- an older research in diagnostics of occupational asbestosis with newer visualization technique.

The first topic presents a new field of research for our team. Some preliminary results are presented as an example of the multiresolution approach.

The second topic is an example how visualization technique can help in diagnosing a disease. It points to the areas with suspicious shadows, which makes a medical doctor to examine these areas.

I. Vujović, *Multiresolution Approach to Processing Images for Different Applications*, SpringerBriefs in Electrical and Computer Engineering,
DOI 10.1007/978-3-319-14457-3_6

6.1 Multiresolution Approach to EMG Analysis

This research is performed with the collaboration of School of Medicine in Split. Stutterers (experimental subjects with problem of stutter) have a speech dysfunction. Our goal is to analyze signals obtained by electrodes attached to a head surface at laryngeal part of human brains. Several signals are recorded during experiments and stored in the edf format.

The source format includes:

- EMG activity of laryngeal sector of brains,
- microphone signal,
- screen sensor, and
- electrical impulse stimuli of brains.

Multiresolution approach is used to correlate laryngeal timings with electrical stimulielectrical stimuli. As an example, two level wavelet decomposition of one measurement is shown in Fig. 6.1. The upper graph shows time sequence of the recorded laryngeal signal. It is analyzed by Daubechies wavelet of 8th order. Since it is not necessary to obtain on-line, high speed, of execution, it is chosen to use higher orders of wavelets. The second reason for using higher orders of wavelets is in the shape of the source signal and the dynamics of the sourced signal. At the first level of the decomposition, the source signal is split into low and high frequency bands. We expect that important phenomena occur in the high frequency band (red ellipsoid and arrow in the middle row of the image).

The third row of the Fig. 6.1 shows the second level of the decomposition. Both the lower and the higher frequency bands are again split into low and high subbands. The decomposition can go further. Intuitive analysis leads to even 32 levels of the decomposition with wavelet orders of even far as 44 to obtain satisfactory and reliable results.

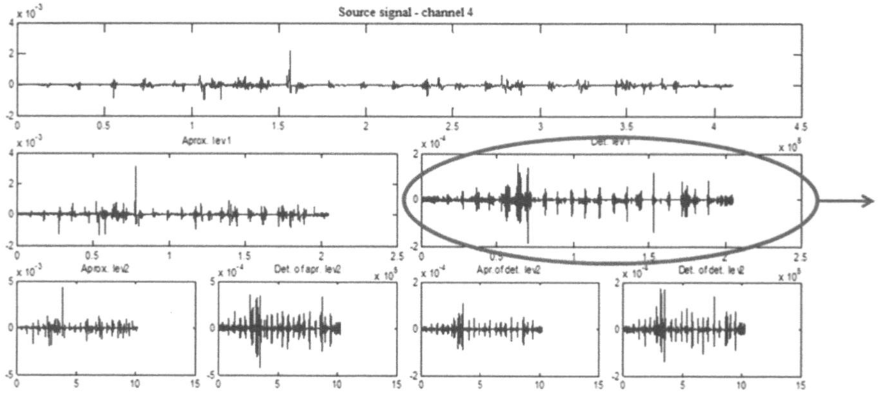

Fig. 6.1 Multiresolution analysis of the EMG signal of the test subject 2

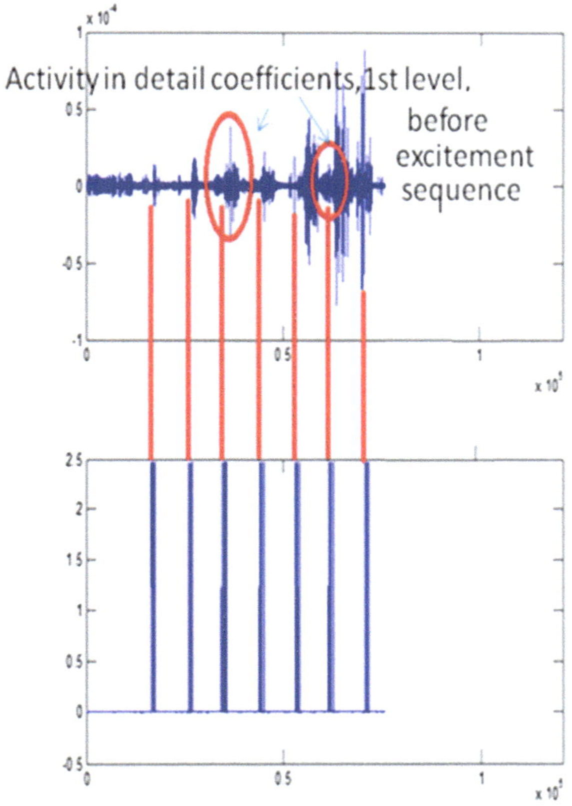

Fig. 6.2 Time synchronization between extinction and the "impulse" response

Figure 6.2 shows time synchronization between the impulse (stimulus) and the respond of the brains. The details coefficients were analyzed. The brains' respond is timed after the electrical stimuli, which is expected, at first. Test subject are expected to talk (to name the image shown to them) after the stimuli. It can be seen that the brains' activity increases after a few stimuli phases, which could be related to the subject's preparations to talk.

It can be seen that WT shows activity in brain before people actually talk. It is our hope that the stutter could be predicted and, perhaps, one day, filtered or reduce by some sort of bleeding-edge electronic devices.

It should be outlined that this is only a preliminary research and that there is no guarantee what will happen with the research in future.

Figure 6.3 shows the spectrograms of the lower and the higher frequency bands. The y-axis represents time and the x-axis the normalized frequency. It can be seen from Fig. 6.3b that the most frequent details coefficients are in the mean of time range, where a half of the frequency range is used. The approximation coefficients (see Fig. 6.3a) do not produce so clearly readable spectrogram. This is

(a)
(b)
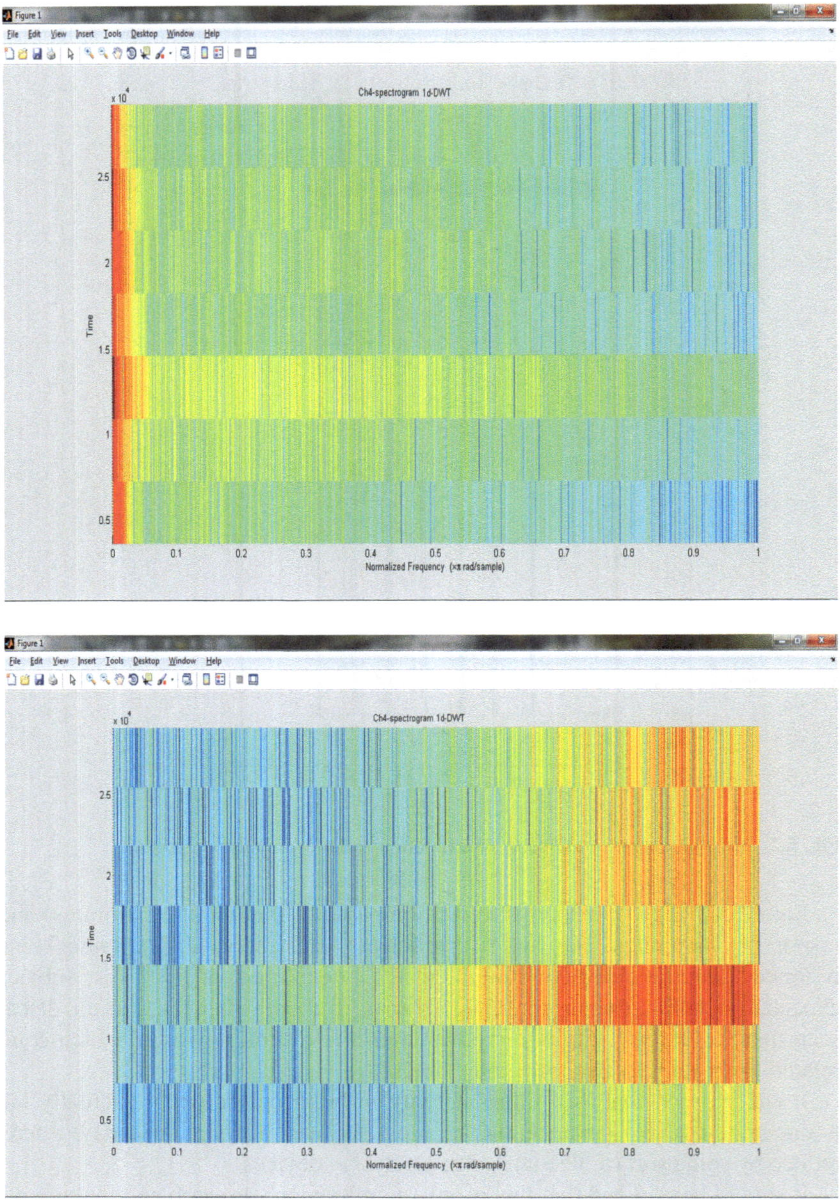

Fig. 6.3 Spectrogram of the analyzed signal: **a** lower frequency band (wavelet approximation coefficients), **b** higher frequency band (wavelet details coefficients)

one of reasons for the higher bands usage—to make conclusions easier. It should be noted the choice of the bands is not general and depend on the application and specifics of the analyzed signals.

The multiresolution approach is used in this preliminary research. It means that a path to the results is chosen from level to level. For example, it is possible to choose HF band at the first level of the decomposition, the LF band at the second, or any other combination. This leads to future tree analysis for the optimal and the reliably results.

In this research, we examine a lot of data (low level processing) to obtain the conclusions (high level information).

6.2 Computer-Aided Diagnostic of Occupational Asbestosis

In this section, a multiresolution approach is presented in cases of image compression and visualization.

Image compression is performed in order to store X-ray images for future use. It was imperative that the medical information is preserved.

Visualization is performed to help medical doctors in a diagnostic procedure. In this case, to point out to the ROIs.

6.2.1 Pulmonary X-Ray Compression

Test subject were asbestos infected patients. Particularly, pulmonary X-ray images were digitalized and input into computer. A multiresolution algorithm was developed for image compression purposes. A goal was to determine maximum compression ratio for reliable medical diagnosis. Two parameters were taken into account: wavelet type and levels of decomposition. Decomposed image coefficients were recorded. The reconstruction had to be performed to obtain the original image size necessary for diagnostics. However, it is possible to observe that an image can be a little fuzzier in some cases after reconstruction. For example, Haar wavelet at 14th level of decomposition is reconstructed to the original size with observable defects, but details (which are shadows in the infected tissue) are so observable that medical diagnosis remains the same as from the original image.

Conclusions of this research were [59]:

- symlets can be used for up to do 14th level of decomposition and obtained $CR = 1316$,
- biorthogonal wavelets cannot be used always. Their reliability is till 10th level of decomposition and $CR = 131.65$,
- reverse biorthogonal wavelets can be used till 14th level of decomposition, because there are of medical value when asbestosis is considered,
- algorithm produce 100 % correct restored images (in sense of medical diagnosis) if mentioned wavelets are used for determined levels of decomposition, and
- memory space can be reduced from 4.5 MB to 3.5 kB under abovementioned conditions (wavelet family and level).

Hence, it is possible to conclude that low level processing (image compression operations) can influence medical diagnostics (high level operation). This is a fine example of clear influence. If a low level algorithm does not perform within desired parameters, than the medical diagnosis would be wrong. If the algorithm operates satisfactory, than the medical conclusion would be correct. It can be concluded that it is possible to obtain high CR with preservation of the medical information.

These conclusions were obtained by help of medical doctors and radiologists [59]. In cases where there was no consensus between experts, this level of decomposition was discarded, because it is unreliable. In this application, there is no space for errors, because storing of the wrong medical information would be disastrous.

6.2.2 *Visualization of Infected Areas*

The second example is in help to medical doctors. The visualization multiresolution algorithm was developed, which emphasizes the areas of asbestos deposits (actually, changed pulmonary tissue). This algorithm must detect suspicious areas (which could be asbestos), and perform a visualization algorithm to point out these areas and make it visible.

Our research showed that only some tones of gray could be asbestos-infected areas. So, it is necessary to emphasized desired tones with different color, which could be more visible to human eyes.

Figure 6.4 shows an example of the operation of the developed algorithm. Text box at the right side of the figure are added for this book and serves as explanation of the algorithm operation. The same is valid for all lines toward the mentioned text box.

Further development should be at the application level, i.e. change of background color, change of the ROI color, etc.

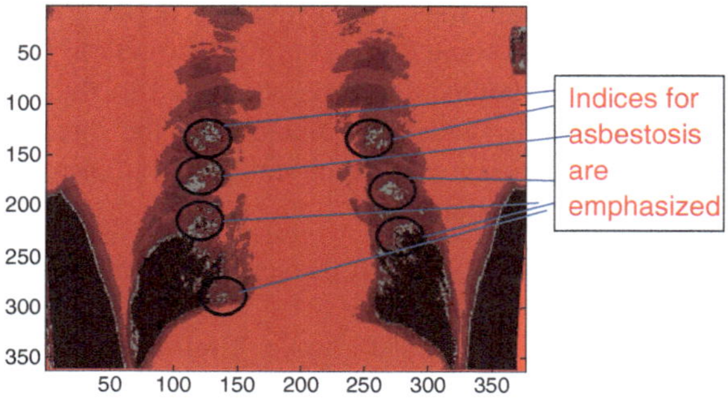

Fig. 6.4 Computer-aided diagnostics: easier detection of shadows, which are indications of asbestosis

Chapter 7
Multiresolution Applications in Video Surveillance

Abstract In this chapter, a multiresolution approach is solving several video surveillance issues is explained. Firstly, edge detection in robot vision is considered. Then, illumination variations problem is addressed with example in people motion detection. Camera jitter problem is addressed in the outdoor example. Removal of weather conditions is considered. Finally, an algorithm for data fusion is presented, which uses wavelet multiresolution approach in oil spill detection from the SAR images.

Keywords Indoor · Outdoor · Robot vision · Edge detection · Camera jitter · SAR images

Surveillance applications can be divided by a space it is covered. If the space is limited, inside buildings, than it is indoor applications. Otherwise, it is outdoor application. This chapter will present out researches from both cases. In this chapter, research in several topics will be presented:

- indoor robot vision,
- indoor motion detection,
- outdoor camera jitter problem,
- influence of weather condition to the problem of traffic surveillance, and
- oil spills detection at open sea.

7.1 Examples of Indoor Applications

The first topic is the robot vision application.

A robot is used indoor, to move between rooms through the corridor. Actual algorithm for motion control is not the scope of our research, but the input to the motion control algorithm. The robot was relatively old and it worked with edge detection to identify its position in the memorized map. Our research was an investigation of different edge detectors performance.

I. Vujović, *Multiresolution Approach to Processing Images for Different Applications*,
SpringerBriefs in Electrical and Computer Engineering,
DOI 10.1007/978-3-319-14457-3_7

(a)

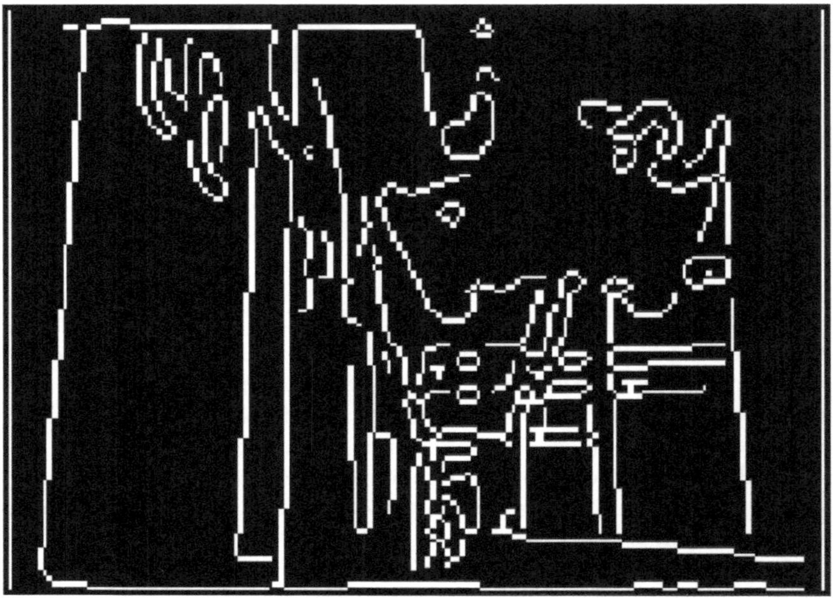

(b)

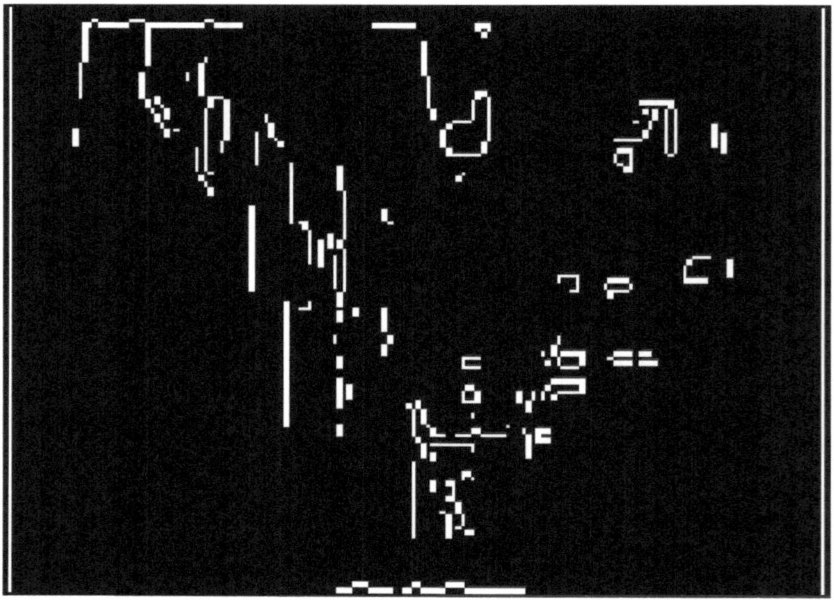

Fig. 7.1 Application of different edge detectors to the same scene: **a** Canny, **b** Perwit, **c** wavelet algorithm

(c)
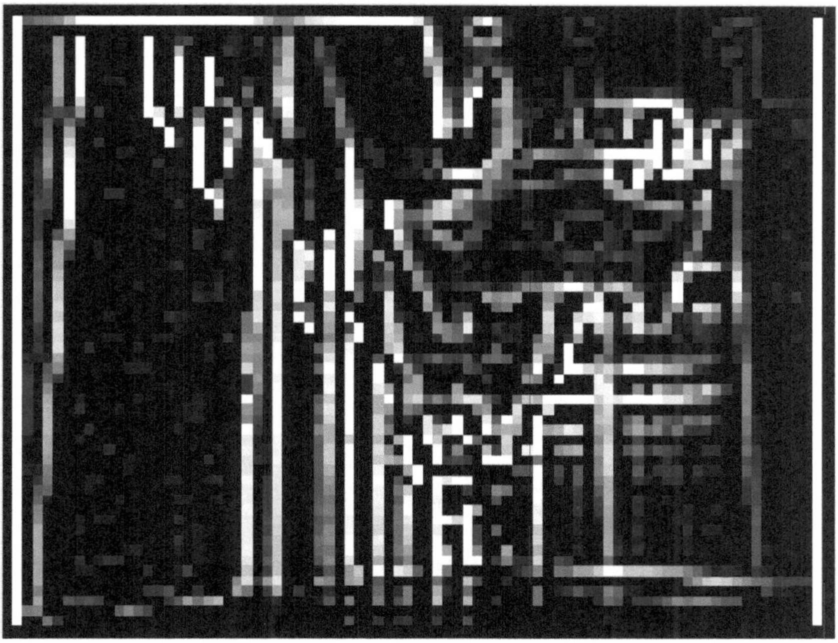

Fig. 7.1 (continued)

Figure 7.1 illustrates a difference between input to the motion control (higher level operation) produced by different edge detectors (low level/vision operation).

It is obvious that different edge detectors (low vision) will influence the data input to the higher vision applications (such as orientation and navigation).

Figure 7.1a shows the result of one of the most popular edge detectors—Canny. Figure 7.1b shows the result for the same frame of the Perwit edge detector. Results of the gradient method applied to the wavelet approximation (1st level of the decomposition) is given (for the same frame) in Fig. 7.1c. The output is not denoised. This is a space for improvement at the low-level stage (in the field of low level image processing). It is not necessary to point out the advantage of better detector to the higher vision applications.

Figure 7.2 shows an example of indoor motion detection in laboratory conditions.

The simple background subtraction algorithm produces an image in Fig. 7.2b. It can be seen that the output image contains a lot of static scene parts, which should be eliminated in the motion mask. Figure 7.2c shows an example of such elimination by the WT.

From Fig. 7.2d it is possible to compare how much of illumination variations were eliminated by the multiresolution approach by the WT.

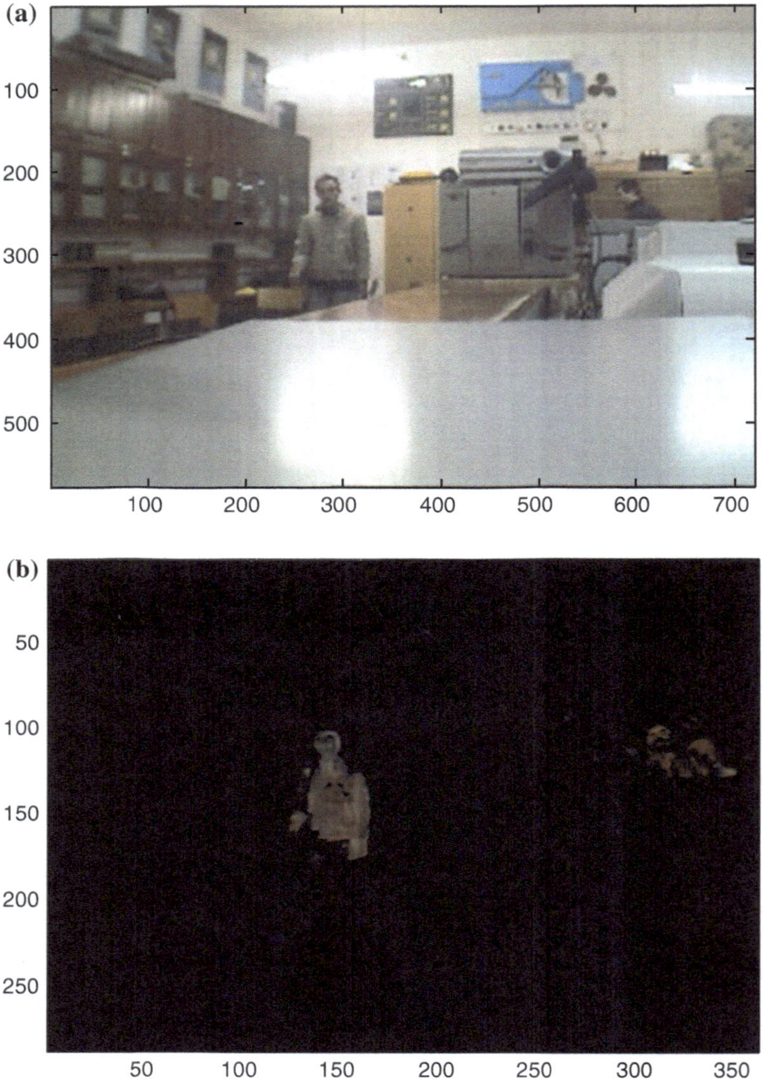

Fig. 7.2 An example of motion detection and related problems in indoor applications: **a** frame 183 in the video sequence, **b** difference in wavelet approximations of the start frame and frame 183, **c** segmented motion, **d** difference of (**c**) and (**b**) for extraction of variations

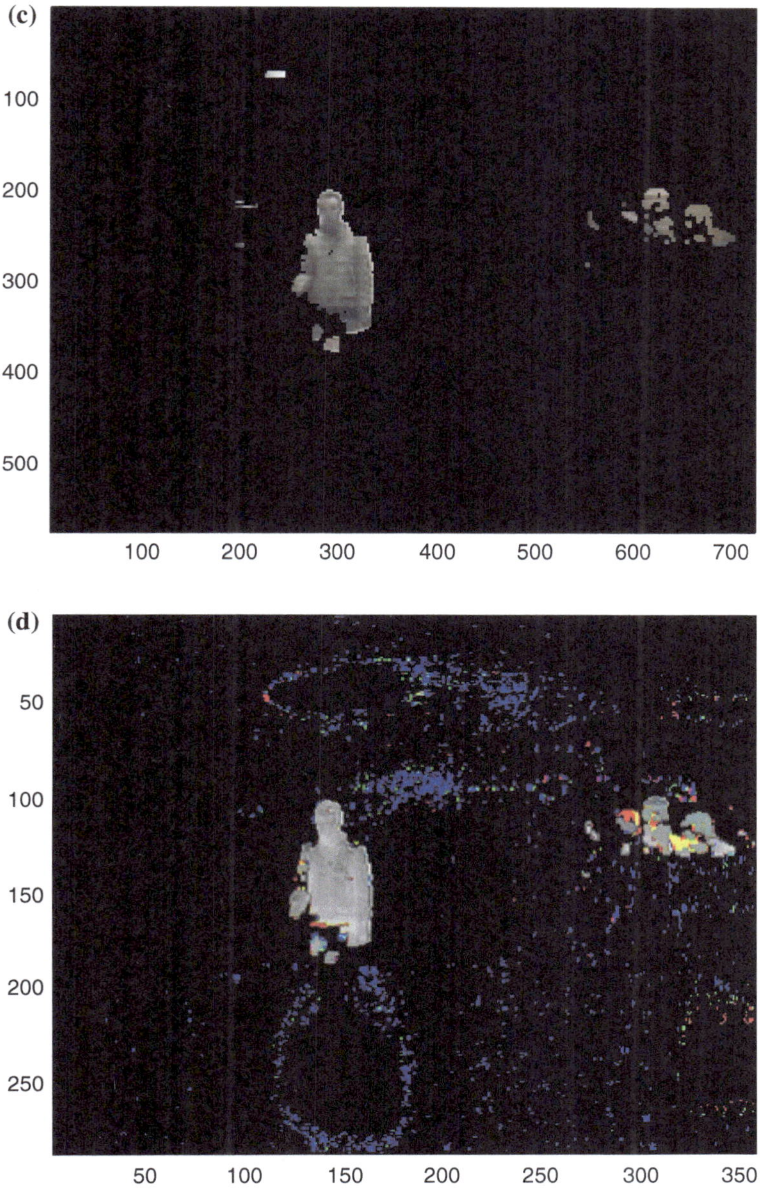

Fig. 7.2 (continued)

7.2 Examples from the Outdoor Applications

One of great problems is the problem of the illumination variations, as illustrated in Fig. 7.3. This problem is present in both the indoor and the outdoor applications of the video surveillance systems. If the considered scene is not in the control environment, there are more problems, such as:

- camera jitter,
- influence of the weather conditions,
- moving shadows,
- part of day (change in Sun's position during daytime),
- waving tree effect, etc.

Figure 7.3 shows advantages of the multiresolution algorithm developed in our research. The developed algorithm is based on [67]. It is improved. The original algorithm [67] is framed as competitive algorithm in Fig. 7.3. The developed algorithm is framed as proposed algorithm in Fig. 7.3. The competitive algorithm is the multiresolution algorithm from the category of memory-based motion detection algorithms. It is based on the energy contained in the coefficients of the lifting wavelet transform (LWT) and the accumulation of it in the buffer.

To make optimization between different algorithm objectives, two wavelets were used. The competitive algorithm uses the 2nd level of decomposition. By implementation of several small changes, the developed algorithm improved robustness to the camera jitter effect. The improvement is confirmed by statistical evaluation using statistical image quality measures—percentage of the correct classifications (PCC) and the precision.

These measures of the image quality are defined with equations:

$$PCC = \frac{TP + TN}{TP + TN + FP + FN} \tag{7.1}$$

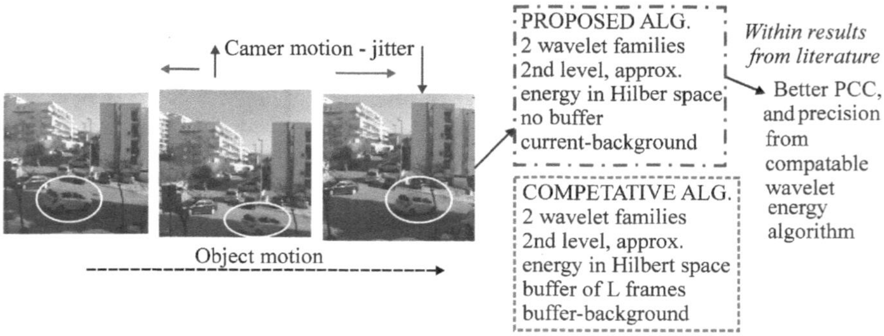

Fig. 7.3 Camera jitter effect reduction by the multiresolution approach

where:

- *TP* is the number of true positive classifications,
- *TN* is the number of true negatives classifications,
- *FP* is the number of false positives classifications, and
- *FN* is the number of false negatives classifications.

It can be said that the PCC measures the absolute "correctness" of the algorithm. PCC defined in (7.1) does not tell everything, because it can be high and that the perceptually the result is not so well. So, the control parameter is introduced. It is *precision*, defined by:

$$precision = \frac{TP}{TP + FP} \tag{7.2}$$

Precision measure is introduced because *FP* is generated by the camera jitter and needs to be minimized to reduce the camera jitter effect.

Another example of problems in the outdoor applications is the weather conditions influence.

An interesting example of the multiresolution approach in solving such a problem was introduced in [68]. Researchers used the method of wavelet multi-level decomposition and wavelet fusion to:

- determine the number of layers of rain and/or snow noise,
- formulate a fusion rule based on rain (snow) noise pollution, and
- make wavelet fusion on specific layer of multiple continuous degraded images.

The first phase of the algorithm is the wavelet decomposition. The images are decomposed for ten layers. This was used to identify the high-frequency coefficients from the 4th layer to the 2nd layer as interested in the detection of rain or snow noise.

The second phase was on the high-frequency coefficients, which were isolated as the rain or snow noise. The coefficient matrix corresponding to each direction of the 4th layer and the 2nd layer is figured out, and the pollution degree *S* matrix corresponding to these coefficients matrix is solved. *S* matrix is made for a unitization processing, which can get the new *S* matrix. The greater the value corresponding to the *S* matrix in the same position is, the more serious the position is polluted by rain or snow noise. This algorithm already uses the local energy. The difference in local energies is expressed with:

$$G = \frac{1}{MN} \sum_{i=1}^{M} \sum_{j=1}^{N} \sqrt{\Delta xf(i,j)^2 - \Delta yf(i,j)^2} \tag{7.3}$$

while the energy of the pixel at the position *i* and *j* is expressed with:

$$E = \frac{1}{MN} \sum_{i=1}^{M} \sum_{j=1}^{N} f(i,j)^2 \tag{7.4}$$

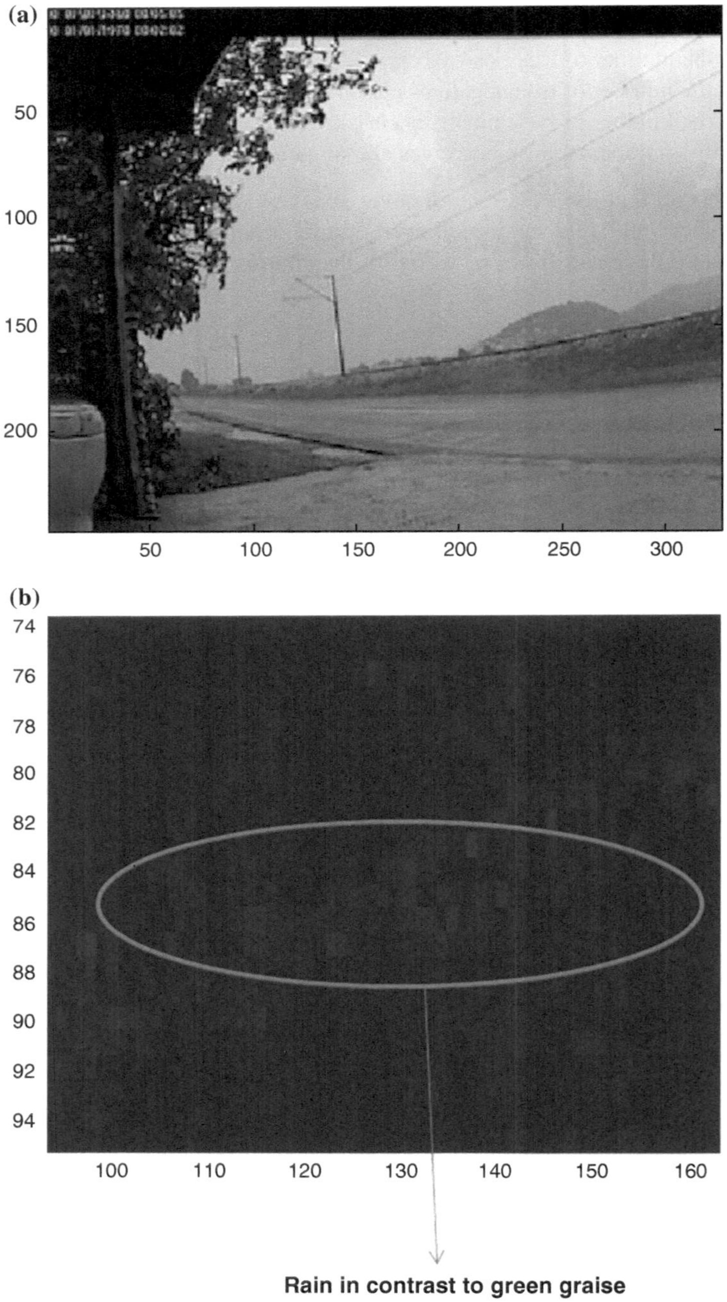

Fig. 7.4 An example of weather conditions and weaving tree effect to the motion detection in the outdoor applications: **a** the original frame, **b** rain as false motion, **c** waving tree effect, **d** rain suppression by the multiresolution approach

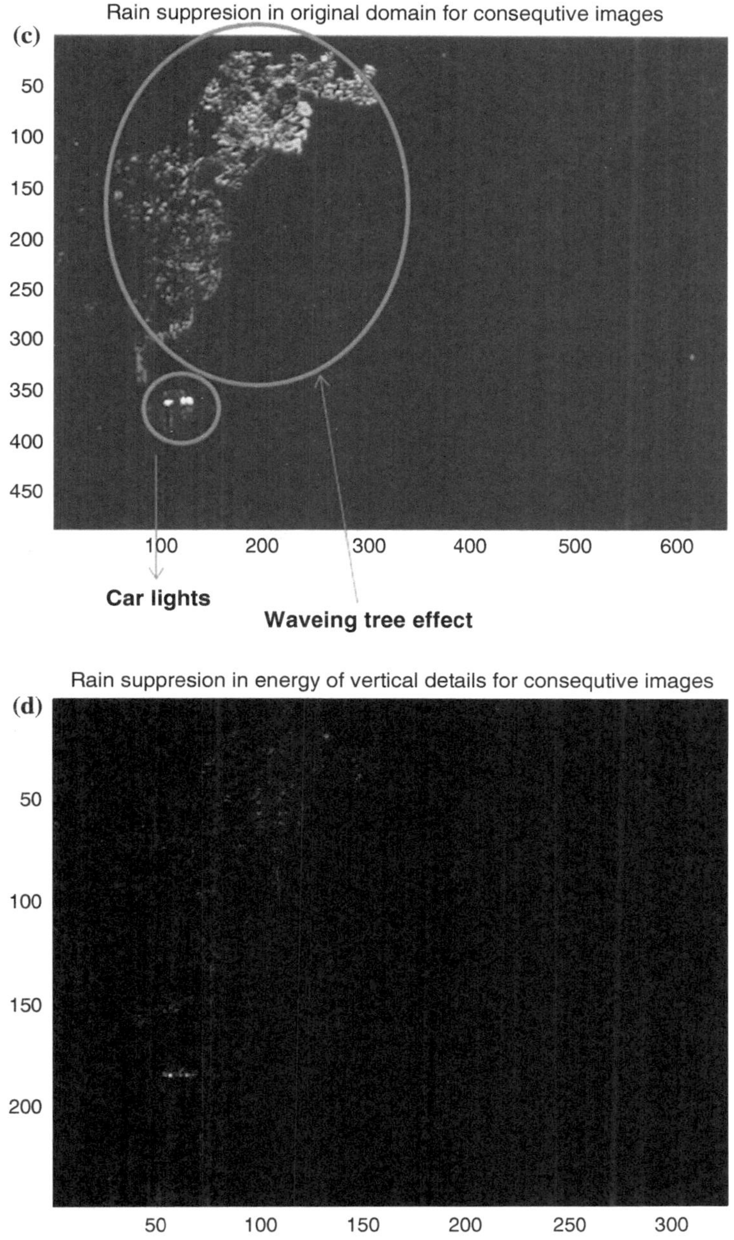

Fig. 7.4 (continued)

Multiplying G and E, we obtain the snow or the rain pollution of the current frame from the video sequence acquired by the video surveillance system:

$$S = G \cdot E \tag{7.5}$$

Figure 7.4 illustrates the abovementioned problems in our research of the road traffic surveillance.

Figure 7.4a shows an example of the current frame's approximation coefficients obtained by the surveillance camera located at the intersection of the state road.

After the first phase of the decomposition, the matrix for the analysis has a half of the elements of the original frame. This matrix of detailed coefficients is visualized in Fig. 7.4b. It can be noticed that the rain drops are detected as the motion, which is not desirable. This effect should be reduced as much as possible. The rain is confused to a motion, because of the contrast between the rain drops and the graise.

Figure 7.4c shows the motion detection result based on the background subtraction without any suppression algorithm. It can be seen that the effect of the waving tree makes almost impossible to distinguish a real motion (a moving car) from the false motion (moving tree due to wind). It happens because the tree branches have a real motion, although it is disturbance from the point of view of our application.

Figure 7.4d shows the results of the rain suppression algorithm to the motion detection. It can be seen that the waving tree effect is also suppressed.

These results are preliminary and there is no guarantee that the conclusion made today will be valid in future.

7.3 SAR Images Analysis for Control of Ecological Incidents

Another research field is the SAR satellite images, which are hard to process without problems at low level processing, such as:

- jitter,
- scatter,
- illumination,
- weather or other problems.

It is easily to find influences of bad threshold determination, for example, to oil spills detection.

General advantages of the multiresolution approach are:

- better filter coefficients,
- faster implementation by SGW,
- filter coefficients obtained by the WT produces better results by the PCC criterion than other techniques, i.e. Chebyshev, Butterworth and ecliptic,
- WT converge faster toward solution thanks to the second generation wavelets.

The algorithm for oil spills detection (Fig. 7.5) developed in our researches is a typical example of the interaction of the low and the high level data. In this case a low level processing is performed at the image processing level with the multiresolution wavelet approach. The data fusion is performed by using higher level information—presence of the ship in the searched area.

The algorithm from Fig. 7.5 generates the alarm based on two types of data. The results are also preliminary and there is no guarantee that the conclusions made up to date will hold in future.

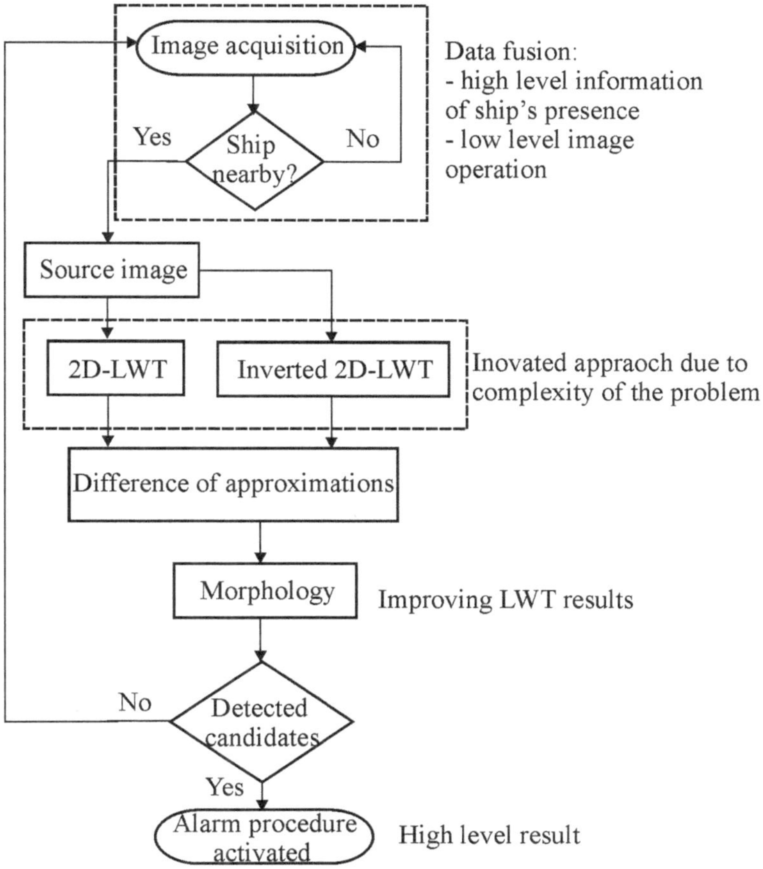

Fig. 7.5 An algorithm for the oil spills detection by data fusion and the multiresolution approach from the satellite SAR images

Chapter 8
Conclusions

Abstract Area covered with this book includes several well-funded industries: the industry of medical electronics, the military industry and the entertainment industry. This chapter summarizes the conclusions from all chapters. General conclusion is the interaction between low-level vision applications and high-level vision applications, which are responsible for the final result of the system.

Keywords Medical electronic industry · Military industry · Entertainment industry · Visual system performance

The area of image processing and analysis, scientific visualization and augmented/virtual reality is of interest to the richest industries in the world, e.g. biomedical or military/security industries. Advances in the field of augmented/virtual reality also include a well-funded industry—entertainment. Industry managers are only interested in final results, and are often not even aware of the existence of problems—lower image processing tasks which greatly influence the performance of higher vision applications.

The influence of lower vision applications on the performance of higher vision applications is illustrated by examples. It is an intuitively interesting topic. The right question should be how to suppress all the possible influences and simultaneously maintain quality and speed of execution of higher vision applications. If we want, for example, to implement an advanced virtual reality based communication system to enable seafarers to communicate with their families, we need to pack and transfer data to maintain resolution and transfer speed. Such applications have to ensure the unpacking of a small number of data through a satellite-band communication channel into a great number of detailed high resolution data. One of the possible solutions is to use the multiresolution approach for data compression. Nowadays, WT is a standard multiresolution tool. One of the ways to improve such applications is to study WT improvements and eventually, find new transformations best suited for the task.

I. Vujović, *Multiresolution Approach to Processing Images for Different Applications*,
SpringerBriefs in Electrical and Computer Engineering,
DOI 10.1007/978-3-319-14457-3_8

The possibilities of development for this field of research are tremendous. We can hope that the newly developed techniques won't be misused.

As stated in the Introduction, research fields presented in this book are:

1. video surveillance,
2. biomedical applications,
3. improved communications through teleoperation, telemedicine, animation, augmented/virtual reality, and robot vision,
4. monitoring of the condition of ship systems and image quality control.

Chapter 4 presented previous research in visual quality control and preliminary results of study of properties of dielectric materials. We arrived at the conclusion that the multiresolution approach is quite promising in these applications.

Although Chap. 5 presented preliminary research results from another field, it also involved the application of the multiresolution approach. It investigated the issue of application of virtual reality in security and contra-terrorist surveillance. The future communication trends and possibilities of the multiresolution approach were presented. Via the data processing part, multiresolution is included between the sensors and the control application, and in obtaining higher resolution in the focused part of the virtual/augmented/animated world and lower resolution in the part of the world marked as unimportant background. The presentation of advantages of such systems in the normal working days of seafarers was of special interest.

Chapter 6 presented preliminary research in EMG signal analysis. Multiresolution approach in CAD was presented in separate sections. CAD was covered by two applications: image compression with the reliable preservation of medical information, and assistance with the diagnostics procedure. The level of image compression without damage to useful (medical) information and the manner of use of different wavelets were established. Furthermore, it was shown that WT exhibits activity in the brain before people actually talk.

Chapter 7 presented the multiresolution approach in video surveillance. Examples of indoor and outdoor applications were presented. It was concluded that the multiresolution approach has some advantages in these applications. The developed algorithm was compared with the competitive algorithm of the same class for the camera jitter problem. The conclusion was that the proposed algorithm had some advantages judging by statistical image quality measurements. Applications covered in this chapter were: the influence of edge detectors on robot vision application in indoor environment, variation illuminations in motion detection for indoor applications, outdoor solution for the camera jitter problem, outdoor rain removal in traffic surveillance, and oil spills detection at open sea by SAR images. The advantages of the multiresolution approach were illustrated by case studies.

References

1. Sundararajan K (2001) Unified point-edgelet feature tracking. MSc thesis, Graduate School of Clemson University. http://www.ces.clemson.edu/~stb/students/ssundar_thesis.pdf. Accessed 15 June 2014
2. Huo X (1999) Combined image representation using edgelets and wavelets. In: 2nd IDR Workshop, Florham Park, 15–17 October 1999. http://www.waveletidr.org/workshop2/posters/xiaoming_huo2.pdf. Accessed 15 May 2014
3. Widynski N, Mignotte M (2011) A contrario edge detection with edgelets. In: 2nd IEEE international conference on signal & image processing applications (ICSIPA'11), Kuala Lumpur, Malaysia, November 2011, pp 421–426
4. Shapelets web pages (2011) http://www.astro.caltech.edu/~rjm/shapelets/. Accessed 12 Feb 2012
5. Bergé J, Réfrégier A, Massey R, Pacaud F, Pierre M et al (2007) Shapelets. In: STEP3 and CFHTLS, STEP workshop, 20 August 2007. http://www.roe.ac.uk/~heymans/step_talks_07/Berge.pdf. Accessed 23 June 2013
6. Ye L, Keogh E (2009) Time series shapelets: a new primitive for data mining. In: 15th ACM SIGKDD international conference on knowledge discovery and data mining (KDD), Paris 2009, pp 947–956. http://alumni.cs.ucr.edu/~lexiangy/Shapelet/kdd2009shapelet.pdf. Accessed 23 April 2014
7. Candés EJ (2014) What is… a curvelet? http://www.ams.org/notices/200311/what-is.pdf. Accessed 12 May 2014
8. Candés E, Demanet L, Donoho D, Ying L (2006) Fast discrete curvelet transforms. Multiscale Model Simul 5(3):861–899. http://authors.library.caltech.edu/6810/2/CANmms06preprint.pdf, http://dx.doi.org/10.1137/05064182X
9. Candès E (2006) Lecture 2: curvelets. http://www.cs.tut.fi/~karen/CandesCurvelets.pdf. Accessed 12 April 2013
10. Blanco-Silva F, Lucier BJ (2006) Curvelets vs wavelets mathematical models of natural images. http://www.math.sc.edu/~blanco/Curvelets/talk.pdf. Accessed 1 March 2013
11. Ma J, Plonka G (2009) A review of curvelets and recent applications. https://www.uni-due.de/~hm0029/pdfs/CurveletReviewshort.pdf. Accessed 1 April 2012
12. Ying L, Demanet L, Candès E (2005) 3D Discrete curvelet transform. In: Proceedings SPIE wavelets XI conference, San Diego, July 2005. http://math.mit.edu/icg/papers/3dFDCT.pdf. Accessed 13 April 2012
13. Candés EJ, Donoho DL (2004) New tight frames of curvelets and optimal representations of objects with C^2 singularities. Comm Pure Appl Math 57(2):219–266. doi:10.1002/cpa.10116, http://statweb.stanford.edu/~candes/papers/CurveEdges.pdf
14. Do MN, Vetterli M (2005) The contourlet transform: an efficient directional multiresolution image representation. IEEE Trans Image Process 14(12):2091–2106

I. Vujović, *Multiresolution Approach to Processing Images for Different Applications*,
SpringerBriefs in Electrical and Computer Engineering,
DOI 10.1007/978-3-319-14457-3

15. Do MN, Vetterli M (2002) Contourlets: a directional multiresolution image representation. In: Proceedings of IEEE international conference on image processing (ICIP), Rochester, September 2002. http://www.ifp.illinois.edu/~minhdo/publications/icip_contourlet.pdf
16. Tsai WS (2008) Contourlet transforms for feature detection. Final project report, University of Texas. http://users.ece.utexas.edu/~bevans/courses/ee381k/projects/spring08/tsai/FinalProjectReport.pdf
17. Reddy TK, Kumaravel N (2011) A comparison of wavelet, curvelet and contourlet based texture classification algorithms for characterization of bone quality in dental CT. In: 2011 international conference on environmental, biomedical and biotechnology IPCBEE, IACSIT Press, Singapore, pp 60–65
18. Jansen M, Oonincx P (2005) Second generation wavelets and applications. Springer, London
19. Strang G, Nquyen T (1997) Wavelets and filter banks. Wellesly–Cambridge Press, Boston
20. Vetterli M, Gall DL (1989) Perfect reconstruction FIR filter banks: some properties and factorizations. IEEE Trans Acoust Speech Signal Process 37(7):1057–1071
21. Vetterli M, Kovačević J (1995) Wavelets and subband coding. Prentice Hall, London
22. Sheng Y (2000) Wavelet transform. In: Poularikas AD (ed) The transforms and applications handbook, 2nd edn. CRC Press LLC, Boca Raton. http://dsp-book.narod.ru/TAH/ch10.pdf
23. Vithalani CH (2012) Lifting scheme of wavelet transform. http://shodhganga.inflibnet.ac.in/bitstream/10603/4341/7/07_chapter%203.pdf. Accessed 2 April 2012
24. Swedelson W (1997) The lifting scheme: a construction of second generation wavelets. SIAM J Math Anal 29(2):511–546
25. Christoper H, Walnut DF (2006) Fundamental papers in wavelet theory. Princeton University Press, London
26. Vujović I, Šoda J, Kuzmanić I (2012) Cutting-edge mathematical tools in processing and analysis of signals in marine and navy. Trans Marit Sci 1(1):35–48
27. Mallat S (2009) A wavelet tour of signal processing, 3rd edn. Academic Press, New York
28. Wang C, Sun R (2009) The quality test of wire bonding. Mod Appl Sci 3(12):50–56
29. Giesko T, Mazurkiewicz A, Zbrowski A (2010) Advanced mechatronic system for in-line automated optical inspection of metal parts. Int J Simul Syst Sci Tech 11(5):33–38
30. Hietaniemi R, Varjo S, Hannuksela J (2013) A machine vision based lumber tracing system. In: VISAPP 2013—international conference on computer vision theory and applications, Barcelona, Spain, 21–24 February 2013, pp 98–103
31. Hecker T (2007) Automating quality control in manufacturing systems combining knowledge-based and rule learning Approaches. Master thesis, Otto-Friedrich-Universität Bamberg.
32. Keser T, Hocenski Ž, Hocenski V (2010) Intelligent machine vision system for automated quality control in ceramic tiles industry. Strojarstvo 52(2):105–114
33. Papakostas GA, Karras DA, Mertzios BG, Boutalis YS (2005) An efficient feature extraction methodology for computer vision applications using wavelet compressed Zernike moments. ICGST Int J Graph Vis Image Process SI1:5–15
34. Morlier J, Salom P, Bos F (2006) New image processing tools for structural dynamic monitoring. http://oatao.univ-toulouse.fr/1739/1/Morlier_1739.pdf. Accessed 15 May 2013
35. Thomas J, Jeray J, Kareem A, Bowyer K (2009) Efficacy of damage detection measures from satellite images. In: 11th Americas conference on wind engineering, San Juan, Puerto Rico, 22–26 June 2009. http://www.iawe.org/Proceedings/11ACWE/11ACWE-Thomas.pdf
36. Joo DJ (2012) Damage detection and system identification using a wavelet energy based approach. Ph.D. thesis, Columbia University
37. Rimac-Drlje S, Keller A, Nyarko KE (2005) Self-learning system for surface failure detection. In: 13th European signal processing conference EUSIPCO 2005, Antalya, Turkey, 4–8 September 2005. http://www.eurasip.org/Proceedings/Eusipco/Eusipco2005/defevent/papers/cr1396.pdf

38. Silveira J, Ferreira MJ, Santos C, Martins T (2008) Computer vision techniques applied to the quality control of ceramic plates. http://repositorium.sdum.uminho.pt/bitstream/1822/16574/1/C27-Computer%20Vision%20Techniques%20Applied%20to%20the%20Quality%20Control%20of%20Ceramic%20Plates.pdf. Accessed 17 Mar 2013
39. Dogar A (2010) Integrating 3D quality control function into an automated visual inspection system for manufacturing industry. UPB Sci Bull C 72(2):63–76
40. Badashah SJ, Subbaiah P (2012) Surface roughness prediction with denoising using wavelet filter. Int J Adv Eng TECH 3(2):168–177
41. Kohut P, Holak K, Uhl T, Monitoring of civil engineering structures supported by vision system. In: 6th European workshop on structural health monitoring, Dresden, Germany, 3–6 July 2012
42. Kuzmanić I, Vujović I (2014) Reliability and availability of quality control based on wavelet computer vision. Paper presented at the 8th international conference on advanced computational engineering and experiment, Paris, 30 June–3 July 2014
43. Kuzmanić I, Vujović I, Šoda J (2014) Damage detection in materials based on computer vision wavelet algorithm. In: Öchsner A, Altenback H (eds) Design and computation of modern engineering materials. Springer, Berlin
44. De Paolis LT (2013) Virtual and augmented reality applications. http://www.iaria.org/conferences2013/filesICONS13/keynote%20De%20Paolis.pdf. Accessed 14 May 2014
45. Merians AS, Jack D, Bolan R, Trenaine M, Burdea GC, Adamovich SV, Recce M, Poizner H (2007) Virtual reality—augmented rehabilitation for patients following stroke. Phys Ther 82(9):898–915
46. You S (2005) Augmented reality—linking real and virtual worlds. University of Southern California. http://graphics.usc.edu/~suyay/class/AR.pdf. Accessed 3 May 2014
47. Wagner D, Reitmayr G, Mulloni A, Drummond T et al (2010) Real-time detection and tracking for augmented reality on mobile phones. IEEE Trans Vis Comput Graph 16(3):355–368
48. Pavić I (2012) 3D animation using Matlab. (in Croatian) Seminar work, University of Split, Faculty of Maritime Studies
49. Kuzmanić I, Vujović I, Vujović M (2012) Impact of visualization and communication technologies on the occupational health of mariners. In: 4th international maritime science conference, Split, Croatia, 16–17 June 2012, pp 143–147
50. Clark KB, Chein I, Cook SW (2004) The effects of segregation and the consequences of desegregation a (September 1952) social science statement in the Brown v. board of education of Topeka Supreme Court case. Am Psychol 59(6):495–501
51. Ulven AJ, Omdal KA, Herløvnielsen H, Irgens Å, Dahl E (2007) Seafarers' wives and intermittent husbands social and psychological impact of a subgroup of Norwegian seafarers' work schedule on their families. Int Marit Health 58:115–127
52. Smith C (2010) 3D TV health risks: 7 people warned not to watch 3D TV. http://www.huffingtonpost.com/2010/04/16/3d-tv-health-risks-7-peop_n_540227.html. Accessed 13 June 2014
53. Šoda J, Beroš SM, Kuzmanić I, Vujović I (2013) Discontinuity detection in the vibration signal of turning machines. In: Öchnser A, Altenbach H (eds) Experimental and numerical investigation of advanced materials and structures, advanced structured materials, vol 41. Springer, Heidelberg, pp 27–54. doi:10.1007/978-3-319-00506-5_3
54. Pennec E, Mallat S (2005) Sparse Geometric Image Representations With Bandelets. IEEE Trans Image Process 14:423–438. doi:10.1109/TIP.2005.843753
55. Candés E, Demanet L, Donoho D, Ying L (2006) Fast discrete curvelet transforms. Multiscale Model Simul 5:861–899. doi:10.1137/05064182X
56. Donoho D (1999) Wedgelets: nearly-minimax estimation of edges. Ann Stat 27:859–897. doi:10.1214/aos/1018031261
57. Selesnick IW, Baraniuk RG, Kingsbury NG (2005) The dual-tree complex wavelet transform. IEEE Signal Process Mag 22:123–151. doi:10.1109/MSP.2005.1550194

58. Kuzmanić I, Vujović I (2015) Reliability and availability of quality control based on wavelet computer vision, Springer briefs in electrical and computer engineering. Springer, Heidelberg
59. Vujović I (2004) Application of wavelets in biomedical data processing in the example of X-ray compression of asbestosis infected pateients. MSc thesis, University of Split, Faculty of Electrical Engineering, Mechanical Engineering and Naval Architecture
60. Vujović I, Kulenović Z, Kežić SV (2014) Proposal of new method for dielectric materials selection in ship system applications. Naše More 61:28–32
61. Kulenović Z, Vujović I, Kežić SV (2014) Simulation of important factors' impact in the choice of dielectric material for marine applications. In: Proceedings of 6th International maritime science conference, Solin, Croatia, 28–29 April 2014
62. Gadani DH (2010) Dielectric properties of soils in microwave region. Ph.D. thesis, The Gujarat University
63. Li HM, Ra CH, Zhang G, Yoo WJ (2009) Frequency and temperature dependence of the dielectric properties of a PCB substrate for advanced packaging applications. J Korean Phys Soc 54:1096–1099
64. Cheng YL, Leon KW, Huang JF, Chang WY, Chang YM, Leu J (2014) Effects of moisture on electrical properties and reliability of low dielectric constant materials. Microelectron Eng 114:12–16
65. Kaiser DR, Reinert DJ, Reichert JM, Minella JPG (2010) Dielectric constant obtained from TDR and volumetric moisture of soils in Southern Brazil. Rev Bras Cienc Solo 34:649–658
66. Helhel S, Colak B, Özen S (2007) Measurement of dielectric constant of thin leaves by moisture content at 4 mm band. Prog Electromag Res Lett 7:183–191
67. Vujović I, Šoda J, Kuzamanić I (2013) Stabilising illumination variations in motion detection for surveillance applications. IET Image Process 7:671–678. doi:10.1049/iet-ipr.2013.0169
68. Zhen C, Jihong S (2013) A new algorithm of rain (Snow) removal in video. J Multimedia 8:168–174
69. Vujović I, Kuzmanić I, Vujović M (2006) Algorithm for Combined wavelet quasi-superresolution. In: Proceedings of 5th international symposium communication systems networks and digital signal processing, Patras, Greece, 19–21 July 2006, pp 469–473
70. Vujović I, Kuzmanić I, Kezić D (2006) Wavelet superresolution and quasi-superresolution in robot vision. In: Proceedings 10th international research/expert conference "Trends in the development of machinery and associated technology", Barcelona-Lloret de Mar, Spain, 11–15 September 2006, pp 597–600

Index

I. Vujović, *Multiresolution Approach to Processing Images for Different Applications*,
SpringerBriefs in Electrical and Computer Engineering,
DOI 10.1007/978-3-319-14457-3

24 $$D(s)^{-1}N_\ell(s) = \sum_{\alpha=0}^{k} W_\alpha s^\alpha + \hat{W}(s),$$

where $\forall\alpha = 0 \sim k$ $W_\alpha \in \mathbb{R}^{\nu\times n_i}$, $\hat{W}(\cdot) \in \mathbb{R}_{p,o}(s)^{\nu\times n_i}$ (whence $t \mapsto W(t) := \mathcal{L}^{-1}[\hat{W}](t)$ is a function on $\mathbb{R}$ with $W(t) = 0$ $\forall t < 0$). As a consequence the z-s p-s trajectory of the PMD is given by

25 $$\xi(t) = \sum_{\alpha=0}^{k} W_\alpha p^\alpha u(t) + \int_{0-}^{t} W(t-\tau)u(\tau)\, d\tau \quad \forall t > 0-$$

where the required differentiations in (25) may have to be interpreted in the distribution sense ($p1(t) = \delta(t)$; $p^\alpha 1(t) = \delta^{(\alpha-1)}(t), \cdots$). Note that, since $W(t) = 0$ for $t < 0$ and $u(t) = \theta$ for t in a sufficiently small interval $(-\varepsilon, 0)$, the convolution integral in (25) is zero on $(-\varepsilon, 0)$.

26 By the substitution of the z-s p-s trajectory (25) in the readout map (2) we obtain the <u>zero-state response</u> (z-s response) <u>of the PMD</u> to the input $u(\cdot)$ as

27 $$y(t) = N_r(p)\{\sum_{\alpha=0}^{k} W_\alpha p^\alpha u(t) + 1(t)\int_{0-}^{t} W(t-\tau)u(\tau)d\tau\} + K(p)u(t), \quad \forall t > 0-,$$

where the required differentiations may again have to be made in the distribution sense; the factor $1(t)$ is inserted in (27) to emphasize that, for $t \in (-\varepsilon, 0)$, the convolution integral equals zero.

Call

28 $$\hat{H}(s) = N_r(s)D(s)^{-1}N_\ell(s) + K(s)$$

the <u>transfer function of the PMD</u> $[D, N_\ell, N_r, K]$; then the Laplace transform of the z-s response reads

29 $$\hat{y}(s) = \hat{H}(s)\hat{u}(s).$$

Let us consolidate:

30 <u>Remarks</u>. (a) The z-i p-s trajectory $\xi(\cdot)$ and the z-i response $y(\cdot)$ of a PMD are C^∞-functions (see (9) and (12)). In particular $\xi(0-) = \xi(0) = \xi(0+)$, $y(0-) = y(0) = y(0+)$.

(b) Since the input $u(\cdot)$ is p. suff. diff. with discontinuity set D, the z-s p-s trajectory $\xi(\cdot)$ and z-s response $y(\cdot)$ of a PMD are, in general, distributions with $\mathbb{R}_+$ as support; in fact, these distributions are C^r-functions except, possibly, for points in D. Hence we may have $\xi(0-) \neq \xi(0+)$ and $y(0-) \neq y(0+)$; furthermore, since there may be a distribution at 0 in the RHS of (25) and (27), $\xi(\cdot)$ and $y(\cdot)$ may not be defined at $t = 0$. For example, with $\xi(\cdot)$ and all its derivatives zero at $t = 0-$, for

$$\begin{bmatrix} 1 & p^2 + p \\ 0 & p \end{bmatrix} \begin{bmatrix} \xi_1(t) \\ \xi_2(t) \end{bmatrix} = \begin{bmatrix} 0 \\ 1(t) \end{bmatrix},$$

the z-s p-s tractory is $\xi_1(t) = -1(t) - \delta(t)$; $\xi_2(t) = t1(t)$, for $t > 0-$.

(c) Concerning the z-s p-s trajectory $\xi(\cdot)$ and the z-s response $y(\cdot)$, note that the maps $u(\cdot) \mapsto \xi(\cdot)$ and $u(\cdot) \mapsto y(\cdot)$ are $\mathbb{R}$-linear.

(d) Suppose that $u(\cdot)$ is p. suff. diff., whence, for some $r \in \mathbb{N}$, $u(\cdot)$ is C^r $\forall t \in \mathbb{R}_+ \backslash D$ (where D is the set of discontinuity points of $u(\cdot)$). Then the z-s p-s trajectory $\xi(\cdot)$ is p. suff. diff. and $\xi(\cdot)$ is C^r on $\mathbb{R}_+ \backslash D$ if and only if $D(s)^{-1} N_\ell(s)$ is proper. (The same holds for the z-s response $y(\cdot)$ if and only if $\hat{H}(s)$ is proper.)

32 Using linearity and (9), (12), (25), and (27) the complete pseudo-state trajectory (p-s trajectory) and complete response of the PMD are given by

$$33 \qquad \xi(t) = \psi(t)Px_\xi(0-) + \sum_{\alpha=0}^{k} W_\alpha p^\alpha u(t) + \int_{0-}^{t} W(t-\tau)u(\tau)\, d\tau, \quad \forall t > 0-,$$

respectively:

$$34 \qquad y(t) = N_r(p)\left\{\psi(t)Px_\xi(0-) + \sum_{\alpha=0}^{k} W_\alpha p^\alpha u(t) + \int_{0-}^{t} W(t-\tau)u(\tau)\, d\tau\right\} + K(p)u(t), \quad \forall t > 0-.$$

In view of Remark 30, we wrote "$x_\xi(0-)$" in (33) and (34), rather than "$x_\xi(0)$," which was legitimately used in the z-i case.

35 Remark. Equations (33) and (34) imply the following important observation: For any p. suff. diff. $u(\cdot) : \mathbb{R}_+ \to \mathbb{R}^{n_i}$ satisfying $u^{(j)}(0-) = \theta_{n_i}$, $\forall j = 0, 1, 2, \cdots$ and for any state at 0-, $x_\xi(0-)$, the complete p-s trajectory and the complete response of the PMD are uniquely defined.

3.2.2. Reachability of PMDs

Consider the PMD $\mathcal{D}$:= $[D, N_\ell, N_r, K]$ described by

1 $$D(p)\xi(t) = N_\ell(p)u(t)$$

$$t \geq 0$$

2 $$y(t) = N_r(p)\xi(t) + K(p)u(t)$$

with characteristics given in (3.2.1.1)-(3.2.1.2).

3 Definition. A z-i p-s trajectory $\xi(\cdot)$ of $\mathcal{D}$ is called reachable iff there is an input $u(\cdot)$ of the form $\sum_{\alpha=0}^{m} u_\alpha \delta^{(\alpha)}(t)$ (where $u_\alpha \in \mathbb{R}^{n_i}$ for $\alpha = 0 \sim m$) which produces a z-s p-s trajectory $\xi_1(\cdot)$ such that for $t > 0$, $\xi_1(t) = \xi(t)$; equivalently, iff $\exists$ an input $\hat{u}(s) \in \mathbb{R}[s]^{n_i}$ which produces a z-s p-s trajectory $\hat{\xi}_1(s) \in \mathbb{R}(s)^\nu$ such that

$$\hat{\xi}_1(s) = \hat{\xi}(s) + \hat{\eta}(s),$$

where $\hat{\eta}(s)$ is the polynomial part of $\hat{\xi}_1(s)$. (Consequently, $\hat{\xi}(s) \in \mathbb{R}_{p,o}(s)^\nu$ and $\mathcal{L}^{-1}[\hat{\eta}](t) = \theta_\nu$ $\forall t > 0$.)

Analysis. Consider equation (1). Let $L(\cdot) \in \mathbb{R}[s]^{\nu\times\nu}$ be a g.c.ℓ.d. of (D, N_ℓ): equiv. $\exists$ polynomial matrices $\bar{D}$ and $\bar{N}_\ell$ s.t.

4 $$D = L\bar{D} \qquad N_\ell = L\bar{N}_\ell$$

and

5 $$(\bar{D}, \bar{N}_\ell) \text{ is } \ell\text{.c., equiv. } \mathrm{rk}[\bar{D}(s) \,\vdots\, \bar{N}_\ell(s)] = \nu, \ \forall s \in \mathbb{C}$$

(or equiv. for some $\bar{U}_\ell$, $\bar{V}_\ell \in E(\mathbb{R}[s])$,

6 $$\bar{N}_\ell\bar{U}_\ell + \bar{D}\bar{V}_\ell = I_\nu, \quad \forall s \in \mathbb{C})$$

10 Theorem. The z-i p-s trajectory $\xi(\cdot)$ of $\mathcal{D}$ is reachable

$$\Leftrightarrow$$

11 $$\bar{D}(p)\xi(t) = \theta_\nu \quad \forall t \geq 0.$$

12 Comment. (a) If (D, N_ℓ) is not ℓ.c., then the only z-i p-s trajectories that are reachable are those that satisfy (11). For example, if $\det L(p_0) = 0$ and $\det \bar{D}(p_0) \neq 0$, then by Theorem 10 the z-i p-s trajectory ξ_0 defined by

13 $$\xi_0(t) = ke^{p_0 t} \text{ where } k \in \mathbb{C}^\nu \text{ is s.t. } D(p_0)k = L(p_0)\,\bar{D}(p_0)\,k = \theta_\nu,$$

is not reachable.

(b) The set of reachable z-i p-s trajectories is a subspace of the linear space of z-i p-s trajectories.

14 Proof of Theorem 10

⇒ : By assumption the z-i p-s trajectory $\xi(\cdot)$ is reachable, or equiv. for some $\hat{u} \in \mathbb{R}[s]^{n_i}$

15 $$D(s)\hat{\xi}_1(s) = N_\ell(s)\hat{u}(s),\ \hat{\xi}_1(s) = \hat{\xi}(s) + \hat{\eta}(s),\ \hat{\eta} \in \mathbb{R}[s]^\nu.$$

Now $\xi(\cdot)$ is a z-i p-s trajectory, equiv. $D(p)\xi(t) = \theta_\nu\ \forall t \geq 0$; hence for some polynomial vector $\pi(s) \in \mathbb{R}[s]^\nu$

16 $$D(s)\hat{\xi}(s) = \pi(s) \text{ and } D(s)^{-1}\pi(s) \in \mathbb{R}_{p,o}(s)^\nu.$$

Using (15), (16), and (4), we obtain successively

$$D(s)[D(s)^{-1}\pi(s) + \hat{\eta}(s)] = N_\ell(s)\hat{u}(s)$$

$$\pi(s) = -D(s)\hat{\eta}(s) + N_\ell(s)\hat{u}(s) = L(s)\,[-\bar{D}(s)\hat{\eta}(s) + \bar{N}_\ell(s)\hat{u}(s)].$$

Hence defining

$$\bar{\pi}(s) := -\bar{D}(s)\hat{\eta}(s) + \bar{N}_\ell(s)\hat{u}(s),$$

where $\bar{\pi}(s)$ is a polynomial vector, we have

17 $$\pi(s) = L(s)\bar{\pi}(s),\ \bar{\pi}(s) \in \mathbb{R}[s]^\nu,$$

i.e., the initial conditions polynomial vector $\pi(\cdot)$ of (16) is divisible without remainder by $L(\cdot) \in \mathbb{R}[s]^{\nu\times\nu}$.
By (16), (17) and (4) we obtain now

18 $$\bar{D}(s)\hat{\xi}(s) = \bar{\pi}(s) \quad \text{and} \quad \hat{\xi}(s) = \bar{D}(s)^{-1}\bar{\pi}(s) \in \mathbb{R}_{p,o}(s)^{\nu}$$

Thus $\hat{\xi}(s)$ may be viewed as the response of $\bar{D}(s)^{-1}$, starting from the zero state at $t = 0-$ to the input $\bar{\pi}(s)$; note here that the Laplace transform takes into account values of $\xi(\cdot)$ on $\mathbb{R}_+$. Hence in the time domain, with $\bar{\pi}(s) := \sum_{\alpha=0}^{m} \pi_\alpha s^\alpha$, we have

19 $$\bar{D}(p)\xi(t) = \sum_{\alpha=0}^{m} \pi_\alpha \delta^{(\alpha)}(t) \qquad t \geq 0,$$

where at $t = 0-$ <u>we have set</u> $\xi^{(j)}(0-) = \theta_\nu \; \forall j = 0, 1, 2, \cdots$. Hence

20 $$\bar{D}(p)\xi(t) = \theta_\nu \; \forall t > 0,$$

i.e., $\xi(\cdot)$ is <u>the solution of the differential equation $\bar{D}(p)\xi(t) = \theta_\nu$ $t \geq 0$ on $(0, \infty)$</u>. Now the original function $t \mapsto \xi(t)$ is defined upon $(0-, \infty)$ and is a <u>C^∞-extension</u> of its restriction on $(0, \infty)$. So for this $\xi(\cdot)$ we must have

11 $$\bar{D}(p)\xi(t) = \theta_\nu \; \forall t \geq 0.$$ ■

⇐ : By assumption (11) holds; hence $\exists \; \bar{\pi}(s) \in \mathbb{R}[s]^\nu$ s.t.

21 $$\bar{D}(s)\hat{\xi}(s) = \bar{\pi}(s).$$

Now, from (6), we have

22 $$\bar{N}_\ell \bar{U}_\ell \bar{\pi} + \bar{D}\,\bar{V}_\ell \bar{\pi} = \bar{\pi}.$$

Also letting $\hat{u} := \bar{U}_\ell \bar{\pi}$, $\hat{\eta} := -\bar{V}_\ell \bar{\pi}$, and noting that (21) implies that

$$D(s)\hat{\xi}(s) = L(s)\bar{\pi}(s),$$

by the use of (4), we rewrite (22) after premultiplication by $L(\cdot)$ as

$$D(s)[\hat{\xi}(s) + \hat{\eta}(s)] = N_\ell(s)\hat{u}(s).$$

Hence by Definition 3 the input $\hat{u} \in \mathbb{R}[s]^{n_i}$ produces, for $t > 0$, the given z-i p-s trajectory $\xi(\cdot)$. (Note that the assumption, $\bar{D}(p)\xi(t) = \theta_\nu$, $\forall t \geq 0$, implies that $D(p)\xi(t) = L(p)\bar{D}(p)\xi(t) = \theta_\nu$, $\forall t \geq 0$.) ∎

From Theorem 10 we have now the following important theorem.

25 Theorem [Complete reachability of a PMD]. Consider the PMD $\mathcal{D}$ described by (1)-(2).
U.t.c.

26 Every z-i p-s trajectory $\xi(\cdot)$ of $\mathcal{D}$ is reachable

⇔

27 (D, N_ℓ) is ℓ.c.

28 Comment. Let L be a g.c.ℓ.d. of (D, N_ℓ) (see equations (4) and (5)). The meaning of Theorems 25 and 10 is that iff L is nonunimodular (equiv. (D, N_ℓ) is not ℓ.c.), then $\exists$ a z-i p-s trajectory of $\mathcal{D}$ which is not reachable or equiv. $\exists$ a solution $\xi(\cdot)$ of $D(p)\xi(t) = \theta_\nu$ $t \geq 0$ which is not a solution of $\bar{D}(p)\xi(t) = \theta_\nu$ $t \geq 0$: "some z-i p-s trajectory cannot be excited by means of an input": "there occurs a loss of nontrivial dynamics when (D, N_ℓ) has a nonunimodular common left factor and we must work in the zero-state input-excitation mode, $\hat{\xi}(s) = D(s)^{-1}N_\ell(s)\hat{u}(s)$." We obtain a dynamical interpretation of the common nonunimodular left factor L in the left fraction $D^{-1}N_\ell$ of the transfer function $\hat{H} = N_r D^{-1} N_\ell + K$ of the PMD.

29 Proof of Theorem 25. Because of (4)-(5), Theorem 10, and the definition of a z-i p-s trajectory, we are done if we can show that, with $L(\cdot)$ a g.c.ℓ.d. of (D, N_ℓ) and with X and $\bar{X}$ the solution spaces of

$$D(p)\xi(t) = \theta_\nu \quad t \geq 0 \quad \text{resp.} \quad \bar{D}(p)\bar{\xi}(t) = \theta_\nu \quad t \geq 0,$$

$L(\cdot)$ is unimodular

⇔

$$\bar{X} = X$$

⇒ : By (4) $D(p) = L(p)\bar{D}(p)$, where $L(\cdot) \in \mathbb{R}[p]^{\nu\times\nu}$; hence $\bar{X} \subset X$. Now by assumption $L(\cdot)$ is unimodular, such that $\bar{D}(p) = L(p)^{-1}D(p)$ with $L(\cdot)^{-1} \in \mathbb{R}[p]^{\nu\times\nu}$; hence $X \subset \bar{X}$.

⇐ : By (4) $D(p) = L(p)\bar{D}(p)$, where $L(\cdot) \in \mathbb{R}[p]^{\nu\times\nu}$. by Assumption $X = \bar{X}$, hence dim X = dim $\bar{X}$. Hence according to Theorem 2.3.5.2, $\partial[\det D] = \partial[\det \bar{D}]$.

Hence $\partial[\det L] = \partial[\det D] - \partial[\det \bar{D}] = 0$. Therefore, L is unimodular. ■

Complete reachability is finally related to the <u>absence of input-decoupling zeros</u> [Ros.1].

33 <u>Definition</u>. $z \in \mathbb{C}$ is called an <u>input-decoupling zero</u> (i-d zero) of the PMD $\mathcal{D}$ described by (1)-(2) iff given any g.c.ℓ.d. $L(\cdot) \in \mathbb{R}[p]^{\nu\times\nu}$ of (D, N_ℓ), $\det L(z) = 0$.

We have then

34 <u>Corollary</u> [Absence of i-d zeros]. Consider the PMD $\mathcal{D}$ described by (1)-(2). U.t.c.

26 Every z-i p-s trajectory $\xi(\cdot)$ of $\mathcal{D}$ is reachable ■

$\Leftrightarrow$

35 $\mathcal{D}$ has no i-d zeros.

36 <u>Exercise</u>. Prove Corollary 34.

37 <u>Exercise</u>. Show that $z \in \mathbb{C}$ is an i-d zero of the PMD $\mathcal{D}$ described by (1)-(2) iff $\text{rk}[D(z) \,\vdots\, N_\ell(z)] < \nu$.

38 <u>Exercise</u>. Show that the SSD [A, B, C, D] is completely controllable iff every z-i p-s trajectory of the PMD [pI - A, B, C, D] is reachable.

3.2.3. <u>Observability of PMDs</u>

We are given a PMD $\mathcal{D} := [D, N_\ell, N_r, K]$ described by

1 $$D(p)\xi(t) = N_\ell(p)u(t)$$

$$t \geq 0$$

2 $$y(t) = N_r(p)\xi(t) + K(p)u(t)$$

with characteristics (3.2.1.1)-(3.2.1.2), and we want to observe the p-s trajectory $\xi(\cdot)$ through the output $y(\cdot)$. Since the contributions of the input $u(\cdot)$ in (3.2.1.34) can be computed separately, we are lead to the following definition.

3 Definition. A z-i p-s trajectory $\xi(\cdot)$ of $\mathcal{D}$ is said to be unobservable iff the corresponding z-i response $y(\cdot)$ satisfies

4 $$y(t) = N_r(p)\xi(t) = \theta_{n_o} \quad \forall t > 0$$

or equivalently, $\xi(\cdot)$ is a solution of the differential equation

5 $$\begin{bmatrix} D(p) \\ \hline N_r(p) \end{bmatrix} \xi(t) = \theta_{\nu+n_o} \qquad \forall t > 0.$$

✖

Analysis. Consider equation (5). Let $R(\cdot) \in \mathbb{R}[s]^{\nu\times\nu}$ be a g.c.r.d. of (N_r, D): equiv. $\exists$ polynomial matrices $\bar{N}_r$ and $\bar{D}$ s.t.

6 $$D = \bar{D}\,R \qquad N_r = \bar{N}_r R$$

and

7 $$(\bar{N}_r, \bar{D}) \text{ is r.c., equiv. } \operatorname{rk}\begin{bmatrix} \bar{D}(s) \\ \hline \bar{N}_r(s) \end{bmatrix} = \nu, \quad \forall s \in \mathbb{C}$$

(or equiv., for some $\bar{U}_r$, $\bar{V}_r \in \mathcal{E}(\mathbb{R}[s])$,

8 $$\bar{U}_r\bar{N}_r + \bar{V}_r\bar{D}_r = I_\nu, \; \forall s \in \mathbb{C}).$$

10 Theorem. A z-i p-s trajectory $\xi(\cdot)$ is unobservable

$\Leftrightarrow$

11 $$R(p)\xi(t) = \theta_\nu \quad \forall t > 0.$$

12 Comment. (a) If (N_r, D) is not r.c., then any g.c.r.d. is not unimodular and according to Fact 2.3.5.50, the differential equation (11) has nontrivial solutions, whence there are nontrivial unobservable z-i p-s trajectories. For example, if $\det R(z_0) = 0$, then by Theorem 10, $t \mapsto k\exp(z_0 t)$ with $R(z_0)k = \theta_\nu$ is an unobservable z-i p-s trajectory.

(b) The set of unobservable z-i p-s trajectories is a subspace of the linear space of z-i p-s trajectories.

13 Proof of Theorem 10

$\Leftarrow$: By assumption $R(p)\xi(t) = \theta_\nu \; \forall t > 0$, hence by (6) $\forall t > 0$

$$\begin{cases} \bar{D}(p)R(p)\xi(t) = D(p)\xi(t) = \theta_\nu \\ \bar{N}_r(p)R(p)\xi(t) = N_r(p)\xi(t) = \theta_{n_o} . \end{cases}$$

Hence $\xi(\cdot)$ is a solution of the differential equation (5), showing that (i) $\xi(\cdot)$ is a z-i p-s trajectory and (ii) $\xi(\cdot)$ is unobservable.

$\Rightarrow$: By assumption $\xi(\cdot)$ is unobservable, hence using (5)-(6) we obtain successively for $t > 0$

$$\theta_{\nu+n_o} = \begin{bmatrix} D(p) \\ \hdashline N_r(p) \end{bmatrix} \xi(t) = \begin{bmatrix} \bar{D}(p) \\ \hdashline \bar{N}_r(p) \end{bmatrix} R(p)\xi(t).$$

Using (8), i.e., premultiplying the last equation by $[\bar{V}_r \,\vdots\, \bar{U}_r]$, we obtain (11). ∎

Theorem 10 leads immediately to the result below; its proof is suggested in Comment 12.

16 Theorem [Complete observability of a PMD]. Consider the PMD $\mathcal{D}$ described by (1)-(2).

U.t.c.

17 Every nontrivial z-i p-s trajectory $\xi(\cdot)$ of $\mathcal{D}$ is observable

$\Leftrightarrow$

18 (N_r, D) is r.c.

19 Comment. Let R be a g.c.r.d. of (N_r, D) (see equations (6)-(7)). The meaning of Theorems 16 and 10 is that iff R is nonunimodular (equiv. (N_r, D) is not r.c.), then $\exists$ a nontrivial unobservable z-i p-s trajectory $\xi(\cdot)$ or equiv. $\exists$ a nontrivial solution $\xi(\cdot)$ of $R(p)\xi(t) = \theta \; \forall t > 0$. "There occurs a loss of nontrivial dynamics when (N_r, D) has a nonunimodular common right factor and we observe only $y(\cdot)$." We obtain a dynamical interpretation for the cancellation of the common nonunimodular right factor R in the right fraction N_rD^{-1} of the transfer function $\hat{H} = N_rD^{-1}N_\ell + K$ of the PMD.

Complete observability is finally related to the absence of output-decoupling zeros [Ros.1].

22 Definition. $z \in \mathbb{C}$ is called an output-decoupling zero (o-d zero) of the PMD $\mathcal{D}$ described by (1)-(2) iff, given any g.c.r.d. $R(\cdot) \in \mathbb{R}[p]^{\nu\times\nu}$ of (N_r, D), $\det R(z) = 0$.

We have then

23 Corollary [Absence of o-d zeros]. Consider the PMD $\mathcal{D}$ described by (1)-(2). U.t.c.

17 Every nontrivial z-i p-s trajectory $\xi(\cdot)$ of $\mathcal{D}$ is observable

$\Leftrightarrow$

24 $\mathcal{D}$ has no o-d zeros.

25 Exercise. Prove Corollary 23.

26 Exercise. Show that $z \in \mathbb{C}$ is an o-d zero of the PMD $\mathcal{D}$ described by (1)-(2) iff $\text{rk}\begin{bmatrix} D(z) \\ \hline N_r(z) \end{bmatrix} < \nu$.

27 Exercise. Show that the SSD [A, B, C, D] is completely observable iff every nontrivial z-i p-s trajectory of the PMD [pI - A, B, C, D] is observable.

3.2.4. Minimality, Hidden Modes, Poles, and Zeros

Consider a PMD $\mathcal{D} := [D, N_\ell, N_r, K]$ described by

1 $$D(p)\xi(t) = N_\ell(p)u(t)$$

$$t \geq 0$$

2 $$y(t) = N_r(p)\xi(t) + K(p)u(t)$$

with characteristics (3.2.1.1)-(3.2.1.2).

3 The PMD $\mathcal{D}$ is said to be minimal iff $\mathcal{D}$ has no decoupling zeros, or equiv. iff $\mathcal{D}$ has neither i-d zeros nor o-d zeros. We say also that $\mathcal{D}$ has no hidden modes.

4 Comment. According to Corollaries 3.2.2.34 and 3.2.3.23, Definition 3 means that every z-i p-s trajectory $\xi(\cdot)$ of $\mathcal{D}$ is reachable and every nontrivial z-i p-s trajectory $\xi(\cdot)$ of $\mathcal{D}$ is observable.

From Theorems 3.2.2.25 and 3.2.3.16 we have also

5 Theorem. The PMD $\mathcal{D}$ described by (1)-(2) is minimal iff (D, N_ℓ) is ℓ.c. and (N_r, D) is r.c. ∎

Minimality also gives the possibility of an easy characterization of poles and zeros of the transfer function $\hat{H}$ of a PMD $\mathcal{D}$. To avoid unnecessary redundancy in the PMD we shall assume that the polynomial matrices $\begin{bmatrix} N_\ell \\ \hdashline K \end{bmatrix}$ and $[N_r \,\vdots\, K]$ have full normal column (resp. row) rank (avoiding "trivial inputs and outputs").

6 Theorem [Poles and zeros]. Consider a minimal PMD $\mathcal{D}$ described by (1)-(2) with transfer function

7 $$\hat{H} = N_r D^{-1} N_\ell + K,$$

where $n := \partial[\det D]$ and

8 $$\mathrm{rk}\begin{bmatrix} N_\ell \\ \hdashline K \end{bmatrix} = n_i, \quad \mathrm{rk}[N_r \,\vdots\, K] = n_o$$

U.t.c.

(a)

9 $$p \in P[\hat{H}] \Leftrightarrow \det D(p) = 0;$$

(b) if

10 $$P(s) = \begin{bmatrix} D(s) & N_\ell(s) \\ \hline -N_r(s) & K(s) \end{bmatrix}$$

denotes the system matrix, then

11 $$z \in Z[\hat{H}] \Leftrightarrow \mathrm{rk}[P(z)] < \nu + \min(n_o, n_i).$$

12 Comments. (a) It follows from criterion (9) and Definition 3.2.1.15 that for a minimal PMD $\mathcal{D}$, $\lambda \in \mathbb{C}$ is an eigenvalue of $\mathcal{D}$ iff λ is a pole of the transfer function $\hat{H}$ of $\mathcal{D}$. For an arbitrary PMD $\mathcal{D}$, an eigenvalue λ of $\mathcal{D}$ is either a decoupling zero or a pole of $\hat{H}$ (see below).

(b) If $\mathcal{D}$ were not minimal and we accept criterion (10), then, according to Excercise 3.2.2.37 and 3.2.3.26, decoupling zeros are zeros of $\hat{H}$. If $\mathcal{D}$ is minimal, this cannot happen: the occurrence of a zero of the transfer function is not associated with unreachability or unobservability of the PMD $\mathcal{D}$.

(c) Assumption (8) bears only on criterion (10).

15 Proof of Theorem 6. Since $\mathcal{D}$ is minimal it follows by Theorem 5 that $D^{-1}N_{\ell} \in \mathbb{R}(s)^{\nu \times n_i}$ is a ℓ.c.f. and $N_r D^{-1} \in \mathbb{R}[s]^{n_o \times \nu}$ is a r.c.f. Therefore, by Theorems 2.4.1.25L and 2.4.1.25R, we can associate with these fractions two generalized Bezout identities having similar properties as those encountered in the beginning of the proof of Theorem 2.4.4.12. The reasoning of this proof yields the desired results.

16 Exercise. Complete the proof of Theorem 6.

17 Classification of the eigenvalues of a PMD. Consider any PMD $\mathcal{D}$ described by (1)-(2). According to Definition 3.2.1.15, $\lambda \in \mathbb{C}$ is an eigenvalue of $\mathcal{D}$ iff $\det D(\lambda) = 0$. According to Definitions 3.2.2.33 and 3.2.3.22, decoupling zeros must be eigenvalues of $\mathcal{D}$. Note also that by successive g.c.d. extractions, e.g., [Kai.1, pp. 580-582], any PMD can be made minimal by canceling left non-unimodular common factors between D and N_{ℓ} and then right nonunimodular common factors between N_r and D. Obviously, this leaves the transfer function $\hat{H}$ unchanged and at the end of this process we get a new PMD $\bar{\mathcal{D}}$ which is minimal and has no decoupling zeros. The cancellation process has removed all decoupling zeros of $\mathcal{D}$. Hence by Theorem 6 and the reasoning above, we have for any PMD $\mathcal{D}$, using ordered lists in which elements are repeated according to their multiplicities in det D:

18 (eigenvalues of $\mathcal{D}$) = (decoupling zeros of $\mathcal{D}$) + (poles of $\hat{H}$),
where $\hat{H}$ is the transfer function of $\mathcal{D}$.

For a complete treatment, see [Kai.1, pp.580-582].

3.3. Well-Formed and Exponentially Stable PMDs

3.3.1. Well-Formed PMDs

In this section we consider a PMD $\mathcal{D} := [D, N_\ell, N_r, K]$ described by (3.2.1.1)-(3.2.1.2) and want to avoid impulsive behavior at t = 0 in the p-s trajectory $\xi(\cdot)$ and response $y(\cdot)$, (a) for every possible value of $\xi(\cdot)$ and its derivatives at t = 0- and (b) for every input $u(\cdot)$ whose Laplace transform $\hat{u}(s) \in \mathbb{R}_{p,o}(s)^{n_i}$ is strictly proper, or equiv. on t > 0 $u(\cdot)$ is a sum of exponential polynomials in t as in (3.2.1.17) (hence $u(\cdot) \in C^\infty$ on t > 0 ($u(\cdot)$ is infinitely differentiable)). By impulsive behavior at t = 0 we mean that the time function contains Dirac terms involving $\delta(\cdot)$, $\delta^{(1)}(\cdot)$, ··· giving rise to a polynomial vector in s when taking the Laplace transform.

Note that by allowing the value of $\xi(\cdot)$ and its derivatives to become arbitrary at t = 0- we are not sure to "catch" a z-i p-s trajectory $\xi(\cdot)$ or a z-i p-s response $y(\cdot)$. To ensure this we introduced in Secs. 2.3.5 and 3.2.1 the notion of state x(0). This will not be done here.

We begin our study by considering differential equations.

3 Preliminary Analysis. Consider the differential equation

4 $$D(p)\xi(t) = \theta,$$

where p = d/dt is the differential operator, $D(\cdot) \in \mathbb{R}[p]^{\nu\times\nu}$ is nonsingular, and $\xi(\cdot)$ is time dependent on $\mathbb{R}_+$. Let

5 $$D(s) := D_0 s^k + D_1 s^{k-1} + \cdots + D_k,$$

where $D_i \in \mathbb{R}^{\nu\times\nu}$, $\forall i$. Solving (4) by the Laplace transform method for arbitrary initial values of $\xi(\cdot)$ and its derivatives at t = 0-, we get the equation

7 $$D(s)\hat{\xi}(s) = \pi(s),$$

where the initial conditions vector $\pi(s) \in \mathbb{R}[s]^\nu$ is given by

8 $$\pi(s) = [s^{k-1}I \,\vdots\, s^{k-2}I \,\vdots\, \cdots \,\vdots\, I]\begin{bmatrix} D_0 & & & 0 \\ D_1 & D_0 & & \\ \vdots & & \ddots & \\ D_{k-1} & \cdots & \cdots & D_0 \end{bmatrix}\begin{bmatrix} \xi(0-) \\ \xi^{(1)}(0-) \\ \vdots \\ \xi^{(k-1)}(0-) \end{bmatrix}$$

and the coefficient matrix on the RHS is called a Toeplitz matrix, e.g., [Kai.1], [Ver.1]. A key feature of the initial conditions vector $\pi(\cdot)$ is that it is a k-dimensional $\mathbb{R}^{\nu}$-linear combination of $\nu\times\nu$ polynomial matrices which are the quotients of successive divisions of $D(s)$ by s^iI, where $i = 1, \cdots, k$. Indeed, we have $\forall i \in \underline{k}$

$$D(s) = s^iI(D_0s^{k-i} + D_1s^{k-i-1} + \cdots + D_{k-i}) + (D_{k-i+1}s^{i-1} + \cdots + D_k).$$

Hence (7)-(8) can be rewritten as

9 $$\hat{\xi}(s) = D(s)^{-1}\pi(s) = [s^{-1}I \,\vdots\, s^{-2}I \,\vdots\, \cdots \,\vdots\, s^{-k}I]\begin{bmatrix} \xi(0-) \\ \xi^{(1)}(0-) \\ \vdots \\ \xi^{(k-1)}(0-) \end{bmatrix}$$

$$-D(s)^{-1}[s^{-1}I \,\vdots\, s^{-2}I \,\vdots\, \cdots \,\vdots\, s^{-k}I]\begin{bmatrix} D_k & D_{k-1} & \cdots & D_1 \\ & D_k & & \vdots \\ & & \ddots & \vdots \\ 0 & & & D_k \end{bmatrix}\begin{bmatrix} \xi(0-) \\ \xi^{(1)}(0-) \\ \vdots \\ \xi^{(k-1)}(0-) \end{bmatrix}.$$

Note that formula (9) gives the solution $\hat{\xi}(s)$ of the differential equation (4) for all initial values of $\xi(\cdot)$ and its derivatives at $t = 0-$. Formula (9) is the key to our main result.

10 Definition. We say that the differential equation (4) is well formed iff for all initial values of $\xi(\cdot)$ and its derivatives at $t = 0-$ the solution $\xi(\cdot)$ does not contain a distribution at $t = 0$ of the form $\sum_{i=0}^{\ell} \xi_i\delta^{(i)}(t)$, where

$\xi_i \in \mathbb{R}^{\nu}$ $\forall i \in \underline{\ell}$, or equiv. in terms of the Laplace transform: $\hat{\xi}(\cdot)$ given by (9) is strictly proper for all values of $\xi(\cdot)$ and its derivatives at $t = 0-$. ¤

11 Comment. Definition 10 guarantees that for all initial values of $\xi(\cdot)$ and its derivatives at $t = 0-$, for $t > 0$ we shall be on a trajectory $\xi(\cdot)$ of the differential equation (4) without impulsive behavior at $t = 0$. Note that we do not guarantee that $\xi(\cdot)$ is continuous at $t = 0$, whence $\xi(0-)$ may be different from $\xi(0+)$ (see Exercises 22 and 23 below).

We have now the following result:

14 Theorem [Well-formed differential equations]. The differential equation (4) is well formed if and only if

15 $$D \in \mathbb{R}[s]^{\nu\times\nu} \text{ has no zeros at } \infty$$

or equiv.

16 $$D^{-1} \in \mathbb{R}_p(s)^{\nu\times\nu}.$$ ■

17 Exercise. Let

$$D(p) = \begin{bmatrix} p & p^3 \\ 0 & p \end{bmatrix}.$$

Clearly, D^{-1} is not proper. Show that for $\xi_1(0-) = \xi_2(0-) = \xi_2^{(2)}(0-) = 0$, the solution of (4) for $t \geq 0$ is given by

$$\xi(t) = \begin{bmatrix} \delta(t)\xi_2^{(1)}(0-) \\ 0 \end{bmatrix},$$

i.e., $\xi(\cdot)$ has impulsive behavior at $t = 0$ $\forall \xi_2^{(1)}(0-) \neq 0$.

18 Exercise [Invariance of well-formedness under exponential weighting]. Consider differential equation (4). Let $a \in \mathbb{R}$ and $\eta(t) := \xi(t)e^{-at}$, whence in terms of the Laplace transform $\hat{\xi}(s+a) = \hat{\eta}(s)$. Let $\tilde{D}(s) := D(s+a)$ $=: \tilde{D}_0 s^k + \tilde{D}_1 s^{k-1} + \cdots + \tilde{D}_k$. Show that

(a) $D(p)\xi(t) = \theta_\nu \quad t \geq 0$

iff

19 $$\tilde{D}(p)\eta(t) = \theta_\nu \quad t \geq 0$$

(b) $\hat{\eta}(s)$ is given by the equation $\tilde{D}(s)\hat{\eta}(s) = \tilde{\pi}(s)$, where $\tilde{\pi}(s)$ is obtained from (8) by replacing D_j and $\xi^{(j)}(0-)$ by resp. $\tilde{D}_j$ and $\eta^{(j)}(0-)$, $\forall j = 0, 1, \cdots, k-1$.

(c) $\hat{\eta}(s)$ satisfies equation (9) by replacing $\pi(s)$, $D(s)$, D_j, $\xi^{(j)}(0-)$ by resp. $\tilde{\pi}(s)$, $\tilde{D}(s)$, $\tilde{D}_j$, $\eta^{(j)}(0-)$, with $j = 1, 2, \cdots, k$.

(d) The differential equation (4) is well formed iff the differential equation (19) is well formed.

20 Proof of Theorem 14. Notice that by Exercise 2.4.4.38, conditions (15) and (16) are equivalent. Now by Definition 10 we must show that (16) holds if and only if $\hat{\xi}(s)$, as given by (9), is strictly proper $\forall \xi^{(j)}(0-)$ for $j = 0, 1, 2, \cdots$.

If: Without loss of generality we may assume that $D(0) = D_k$ is nonsingular. Indeed, if this does not hold, then, since $D(\cdot)$ is nonsingular, $\exists\, a \in \mathbb{R}$ s.t. $D(a)$ is nonsingular. It suffices then to consider $\eta(t) := \xi(t)e^{-at}$, which satisfies (19) with $\tilde{D}(0) = D(a) = \tilde{D}_k$ nonsingular (see Exercise 18). Now, since by assumption, $\forall \xi^{(j)}(0-)$, $j = 0, 1, 2, \cdots$, $\hat{\xi}(s) \in \mathbb{R}_{p,o}(s)^\nu$, we may set $\xi^{(j)}(0-) = \theta \ \forall j = 1, 2, \cdots$, and obtain from (9)

$$D(s)^{-1}s^{-1}D_k\xi(0-) = s^{-1}\xi(0-) - \hat{\xi}(s) \quad \forall \xi(0-).$$

Since the RHS is strictly proper $\forall \xi(0-)$, it follows that the LHS is strictly proper $\forall \xi(0-)$ with D_k nonsingular. Hence $D(s)^{-1}s^{-1}$ is strictly proper, whence (16) must hold.

Only if: By assumption D^{-1} is proper, hence the RHS of (9) is strictly proper $\forall \xi^{(j)}(0-)$, $j = 0, 1, 2, \cdots$. Therefore, the LHS of (9) must have the same property. ∎

21 Remark. Observe that by Exercises 2.4.4.36.R and 2.4.4.36.L, D^{-1} is proper when the nonsingular polynomial matrix D is column- or row-reduced. Observe also that the same applies if D is in upper-triangular Hermite row form. Hence any differential equation (4) where D is such a matrix is well formed.

22 Exercise. Let the differential equation (4) be well formed. With $D(s)$

given by (5), show that

$$\xi(0+) = \xi(0-) - [D(s)^{-1}]_{s=\infty}[D_k\xi(0-) + D_{k-1}\xi^{(1)}(0-) + \cdots + D_1\xi^{(k-1)}(0-)].$$

(Hint: Use the initial value Theorem (e.g., [Kai.1, p. 12]), viz., $\xi(0+) = \lim_{s\to\infty} s\,\hat{\xi}(s)$, and (9)).

23 Exercise. Let the differential equation (4) be well formed. Show that, for all initial values at $t = 0-$ of $\xi(\cdot)$ and all its derivatives, $\xi(\cdot)$ is continuous at $t = 0$ (equiv. $\xi(0-) = \xi(0+)$) if and only if D^{-1} is strictly proper. (Hint: Use the initial value Theorem, and argue as in the proof of Theorem 14.)

24 Exercise. Show that when differential equation (4) is s.t. D^{-1} is strictly proper, then (a) $\xi(0) := \xi(0+) = \xi(0-)$ is a partial state of (4), or equiv., with $n := \partial[\det D]$, there exists a state $x(0) \in \mathbb{R}^n$ (see Definition 2.3.5.16) such that $x(0) = [\xi(0)^T; x_2(0)^T]^T$, and (b) $n \geq \nu$. (Hint: Use $\xi(t) = \Psi(t)x(0)$, where $\Psi(0) : x(0) \to \xi(0)$ is now surjective; hence modulo a coordinate transformation $\bar{x}(0) = Px(0)$, $\det P \neq 0$, we get $\Psi(0)P^{-1} = [I_\nu \; 0]\cdots$).

Consider now a PMD $\mathcal{D} := [D, N_\ell, N_r, K]$ described by (3.2.1.1)-(3.2.1.2).

25 Definition. We say that the PMD $\mathcal{D}$ is zero-input well formed (z-i well formed) iff for all initial values of $\xi(\cdot)$ and its derivatives at $t = 0-$, the z-i p-s trajectory $\xi(\cdot)$ and the z-i response $y(\cdot)$ do not contain a distribution at $t = 0$ of the form $\sum_{i=0}^{\ell} \xi_i\delta^{(i)}(t)$, resp. $\sum_{j=0}^{m} y_j\delta^{(j)}(t)$, where, $\forall i$ and j, $\xi_i \in \mathbb{R}^\nu$ resp. $y_j \in \mathbb{R}^{n_o}$.

26 Analysis. The z-i p-s trajectory $\xi(\cdot)$ and the z-i response $y(\cdot)$ of $\mathcal{D}$ satisfy

27 $$D(p)\xi(t) = \theta_\nu$$

$$\forall t \geq 0,$$

28 $$y(t) = N_r(p)\xi(t)$$

where these equations can be rewritten as a differential equation

28 $$\begin{bmatrix} I & -N_r(p) \\ 0 & D(p) \end{bmatrix} \begin{bmatrix} y(t) \\ \xi(t) \end{bmatrix} = \begin{bmatrix} \theta_{n_o} \\ \theta_\nu \end{bmatrix} \quad \forall t \geq 0.$$

Since obviously the initial values of $y(\cdot)$ and its derivatives at $t = 0-$ are not needed, it follows immediately that the PMD $\mathcal{D}$ is z-i well formed iff the differential equation (28) is well formed. Hence by Theorem 14 and Fact 2.4.4.39, we have

29 Theorem [z-i well formed PMDs]. The PMD $\mathcal{D}$ described by (3.2.1.1) and (3.2.1.2) is z-i well formed if and only if

30 D^{-1} and $N_r D^{-1}$ are proper rational matrices or equivalently

31 $$\partial[\det D] = \delta_M\left(\begin{bmatrix} N_r \\ -- \\ D \end{bmatrix}\right),$$

where as in Fact 2.4.4.39, δ_M denotes the McMillan degree of the polynomial matrix between the braces and equals the highest degree of any minor of any order. ■

From Exercise 2.4.4.36.R and Theorem 2.4.3.25.R we have

32 Corollary. Consider a PMD $\mathcal{D}$ described by (3.2.1.1) and (3.2.1.2). Then $\mathcal{D}$ is z-i well formed if

(a) D is column reduced,

(b) $\forall j \in \underline{\nu}$, $\partial_{cj}[N_r] \leq \partial_{cj}[D]$. ■

A sharper (and more practical) result is, however, still available.

33 Corollary. Consider a PMD $\mathcal{D}$ described by (3.2.1.1) and (3.2.1.2). Let, as in Remark 2.4.3.14, $R \in \mathbb{R}[s]^{\nu\times\nu}$ be any unimodular matrix, obtained by e.c.o.'s on $\begin{bmatrix} D \\ N_r \end{bmatrix}$, s.t.

$$DR = \bar{D} \quad \text{and} \quad N_r R = \bar{N}_r,$$

where $\bar{D}$ is c.r.

U.t.c.

$\mathcal{D}$ is z-i well formed
if and only

(a) $\forall j \in \underline{\nu}$ $\partial_{cj}[R] \leq \partial_{cj}[\bar{D}]$,

(b) $\forall j \in \underline{\nu}$ $\partial_{cj}[\bar{N}_r] \leq \partial_{cj}[\bar{D}]$. ∎

34 Exercise. Prove Corollary 33. (Hint: Use (30), Fact 2.4.4.43.R, and Theorem 2.4.3.25.R.)

35 Definition. We say that PMD $\mathcal{D} = [D, N_\ell, N_r, K]$, described by (3.2.1.1) and (3.2.1.2), is zero-state well formed (z-s well formed) iff every input $u(\cdot)$ with Laplace transform $\hat{u} \in \mathbb{R}_{p,o}(s)^{n_i}$ produces a z-s p-s trajectory $\xi(\cdot)$ and a z-s response $y(\cdot)$ such that their Laplace transforms satisfy $\hat{\xi} \in \mathbb{R}_{p,o}(s)^\nu$ and $\hat{y} \in \mathbb{R}_{p,o}(s)^{n_o}$ (Note that the class of time functions, whose Laplace transform is a strictly proper rational vector, is precisely the class of C^∞-functions on $t > 0$ that are sums of exponential polynomials in t as in (3.2.1.17)).

36 Analysis. Note that for a PMD $\mathcal{D}$, described by (3.2.1.1) and (3.2.1.2), the Laplace transforms of the z-s p-s trajectory $\xi(\cdot)$ and z-s response $y(\cdot)$ are given by

37 $$\hat{\xi} = D^{-1}N_\ell\hat{u}$$

38 $$\hat{y} = \hat{H}\hat{u}$$

where

39 $$\hat{H} = N_r D^{-1} N_\ell + K$$

is the transfer function of the PMD $\mathcal{D}$.

Note also the following dynamical interpretation of a pole at ∞ (see Definition 2.4.4.35).

40 Theorem [Dynamical interpretation of a pole at ∞]. Let $\hat{H} \in \mathbb{R}(s)^{n_o \times n_i}$ be a given transfer function.

U.t.c.

$\hat{H}$ has a pole at ∞

if and only if

there exists an input $u(\cdot)$ whose Laplace transform $\hat{u} \in \mathbb{R}_{p,o}(s)^{n_i}$ which produces an output $\hat{y} = \hat{H}\hat{u} \in \mathbb{R}(s)^{n_o}$ s.t. $\hat{y}$ has a polynomial part; more precisely $\hat{y}$ is not a strictly proper rational vector (equiv. for that input,

$y(\cdot)$ will have a nonzero distribution at $t = 0$).

41 Exercise. Prove Theroem 40. (Hint: Use Exercise 2.4.4.37.)

By (37), (38), Exercise 2.4.4.37, and Theorem 40, we obtain the following result.

42 Theorem [z-s well-formed PMDs]. A PMD $\mathcal{D} = [D, N_r, N_\ell, K]$ described by (3.2.1.1) and (3.2.1.2) is z-s well formed if and only if

43 $$D^{-1}N_\ell \in \mathbb{R}_p(s)^{\nu \times n_i}$$

and

44 $$\hat{H} := N_r D^{-1} N_\ell + K \in \mathbb{R}_p(s)^{n_o \times n_i},$$

where H is the transfer function of the PMD. ∎

We then have, finally:

47 Definition. We say that the PMD $\mathcal{D} = [D, N_\ell, N_r, K]$, described by (3.2.1.1) and (3.2.1.2), is well formed iff for every initial value of $\xi(\cdot)$ and its derivatives at $t = 0-$, and for every input $u(\cdot)$ such that (a) $u^{(j)}(0-) = \theta_{n_i}$ $\forall j = 0, 1, 2, \cdots$, and (b) $\hat{u} \in \mathbb{R}_{p,o}(s)^{n_i}$, we have that the p-s state trajectory $\xi(\cdot)$ and response $y(\cdot)$ of $\mathcal{D}$ satisfy $\hat{\xi} \in \mathbb{R}_{p,o}(s)^{\nu}$ and $\hat{y} \in \mathbb{R}_{p,o}(s)^{n_o}$ (equiv. they must not contain a distribution at $t = 0$).

48 Analysis. Because of linearity of the responses and Definitions 25 and 35: the PMD $\mathcal{D}$ is well formed iff (a) it is z-i well formed and (b) it is z-s well formed. So by Theorems 29 and 42, $\mathcal{D}$ is well formed iff the rational matrices D^{-1}, $N_r D^{-1}$, $D^{-1}N_\ell$, $\hat{H} = N_r D^{-1} N_\ell + K$ are proper. These four matrices are block entries of an extended transfer function $\hat{H}_e$ [Kai.1, pp. 569-570], given by

49 $$\hat{H}_e := \left[\begin{array}{c|c|c} I_{n_o} & N_r D^{-1} & -\hat{H} \\ \hline 0 & D^{-1} & -D^{-1}N_\ell \\ \hline 0 & 0 & I_{n_i} \end{array}\right] .$$

Note also that $\hat{H}_e^{-1}$ is the polynomial matrix

50 $$\hat{H}_e^{-1} = \left[\begin{array}{c|c|c} I_{n_o} & -N_r & K \\ \hline 0 & D & N_\ell \\ \hline 0 & 0 & I_{n_i} \end{array}\right].$$

Hence from (49)-(50) and Exercise 2.4.4.38, the PMD $\mathcal{D}$ is well formed iff $\hat{H}_e^{-1}$ has no zeros at ∞. Note now that the system-matrix P of $\mathcal{D}$ [Ros.1] is given by

51 $$P = \left[\begin{array}{c|c} D & N_\ell \\ \hline -N_r & K \end{array}\right];$$

i.e., modulo a block-row permutation, $\hat{H}_e^{-1}$ is an extended system matrix obtained by bordering the system matrix appropriately by 0 or I. Using Fact 2.4.4.39, we have then that $\mathcal{D}$ is well formed iff $\partial[\det D] = \delta_M[P]$ = the highest degree of any minor of any order of P (exercise).

Hence we have obtained

52 Theorem [Well-formed PMDs]. Consider a PMD $\mathcal{D} := [D, N_r, N_\ell, K]$ described by (3.2.1.1) and (3.2.1.2), with transfer function matrix $\hat{H} \in \mathbb{R}(s)^{n_o \times n_i}$ and system matrix $P \in E(\mathbb{R}[s])$ given by

53 $$\hat{H} = N_r D^{-1} N_\ell + K$$

and

51 $$P = \left[\begin{array}{c|c} D & N_\ell \\ \hline -N_r & K \end{array}\right].$$

U.t.c.

54 $\mathcal{D}$ is well formed

⇔

55 the rational matrices D^{-1}, $N_r D^{-1}$, $D^{-1}N_\ell$, $\hat{H}$ are proper

⇔

56 $$\partial[\det D] = \delta_M[P],$$

where δ_M denotes the McMillan degree of the polynomial matrix P and is the highest degree of any minor of any order of P. ∎

57 Exercise. Consider the scalar case, i.e., a PMD $\mathcal{D} = [d, n_\ell, n_r, k]$, where all elements are polynomials. Check that (56) $\Leftrightarrow$ (55).

58 Comment. Condition (56) is a degree-limiting condition in order that (55) may hold: "Given a D s.t. D^{-1} is proper, there is, roughly speaking, a bound on the degrees of elements of N_r, N_ℓ, and K."

It is difficult to check condition (56). However a sufficient condition for checking condition (55) is readily available. For this we need the notion of a row-column-reduced polynomial matrix.

60 Definition. Let $D \in \mathbb{R}[s]^{\nu\times\nu}$ be a nonsingular polynomial matrix; then D is said to be row-column-reduced (r.c.r.) iff there exist integers $r_i \geq 0$, $i \in \underline{\nu}$, and $k_j \geq 0$, $j \in \underline{\nu}$, s.t.

61 $$\lim_{s\to\infty} \text{diag}[s^{-r_i}]_{i=1}^{\nu}\, D(s)\, \text{diag}[s^{-k_j}]_{j=1}^{\nu} = D_h \in \mathbb{R}^{\nu\times\nu} \text{ with } \det D_h \neq 0.$$

The integers r_i and k_j are called row powers, resp. column powers, and D_h is called the highest degree coefficient matrix.

62 Comment. Note that condition (61) is equivalent to either one of the two following factorizations of $D \in \mathbb{R}[s]^{\nu\times\nu}$:

63 $$D = \text{diag}[s^{r_i}]_{i=1}^{\nu}\, D_-\, \text{diag}[s^{k_j}]_{j=1}^{\nu}$$

or

64 $$D = D_\ell\, D_-\, D_r,$$

where $D_- \in \mathbb{R}(s)^{\nu\times\nu}$ is biproper (equiv. D_- and D_-^{-1} are proper), and $D_\ell \in \mathbb{R}[s]^{\nu\times\nu}$ and $D_r \in \mathbb{R}[s]^{\nu\times\nu}$ are row-reduced with row-degrees r_i, $i \in \underline{\nu}$, resp. column-reduced with column-degrees k_j, $j \in \underline{\nu}$. Moreover, according to Fact 2.4.3.21, a r.r. matrix $D \in \mathbb{R}[s]^{\nu\times\nu}$ is r.c.r. with column powers zero and similarly a c.r. matrix $D \in \mathbb{R}[s]^{\nu\times\nu}$ is r.c.r. with row powers zero. ∎

65 Fact. Let $D \in \mathbb{R}[s]^{\nu\times\nu}$ be a nonsingular polynomial matrix with entries $d_{ij} \in \mathbb{R}[s]$, where $i, j \in \underline{\nu}$.

U.t.c.

66 D is r.c.r. with row powers $r_i \geq 0$, $i \in \underline{\nu}$, column powers $k_j \geq 0$, $j \in \underline{\nu}$, and D_h in (61) diagonal

if and only if

67 $$\forall i \in \underline{\nu} \quad \partial[d_{ii}] = r_i + k_i, \\ \forall i,j \in \underline{\nu}, \text{ with } i \neq j \quad \partial[d_{ij}] < r_i + k_j.$$ ∎

Proof. Use condition (63) with $D_h = D_-(\infty)$ diagonal. ∎

A nonsingular matrix $D \in \mathbb{R}[s]^{\nu\times\nu}$ <u>can be made r.c.r. by e.o.'s over</u> $\mathbb{R}[s]\cdots$.

68 Remark. Any nonsingular matrix $D \in \mathbb{R}[s]^{\nu\times\nu}$ can be made c.r. by the method of Remark 2.4.3.14 by e.c.o.'s. Moreover, column permutations and premultiplication by a nonsingular constant matrix will further make D c.r. with column degrees $c_1 \geq c_2 \geq \cdots \geq c_\nu \geq 0$ and with highest column degree coefficient matrix $\bar{D}_h = I$. For example, $D \in \mathbb{R}[s]^{\nu\times\nu}$ has been converted into

69 $$\bar{D}(s) = \begin{bmatrix} s^5 & s & s \\ s^4 & s^3 & 1 \\ s^2 & s^2 & s^2 \end{bmatrix},$$

where $\nu = 3$, $c_1 = 5$, $c_2 = 3$, $c_3 = 2$, and $\bar{D}_h = I$. It is a fact that any c.r. matrix $\bar{D} \in \mathbb{R}[s]^{\nu\times\nu}$ with column degrees $c_1 \geq c_2 \geq \cdots \geq c_\nu \geq 0$ and $\bar{D}_h = I$ can, by e.c.o.'s, be converted into a r.c.r. matrix $\tilde{D}$ with $\tilde{D}_h = I$ and with row and column powers given by any pair of ν-tuples of integers $r_i \geq 0$, $i \in \underline{\nu}$, resp. $k_j \geq 0$, $j \in \underline{\nu}$ s.t.

70 $$r_1 \geq r_2 \geq \cdots \geq r_\nu \geq 0, \; k_1 \geq k_2 \geq \cdots \geq k_\nu \geq 0, \; r_i + k_i = c_i, \; \forall i \in \underline{\nu}.$$

The method [Ros.2, p. 841, line 19 from below] performs e.c.o.'s on columns $j = 1, 2, \cdots, \nu-1$ of $\bar{D}$; these e.c.o.'s are conditioned by the replacement of an entry (i,j), $i > j$, by its remainder after division by entry (i, i). This is done in such a way that the degrees of entries (i, j), $\forall i > j$, are lowered below those of entries (i, i), while the degrees of entry (j, j) and entries (i, j), $\forall i < j$, are maintained constant resp. below the degree of entry (j, j); moreover, each diagonal entry remains monic. Hence using (70) there results a matrix $\tilde{D}$ s.t.

$$\forall i \quad \partial[\tilde{d}_{ii}] = c_i = r_i + k_i ,$$
$$\forall i > j \quad \partial[\tilde{d}_{ij}] < c_i = r_i + k_i \leq r_i + k_j ,$$
$$\forall i < j \quad \partial[\tilde{d}_{ij}] < c_j = r_j + k_j \leq r_i + k_j .$$

As a result, condition (67) is satisfied and by Fact 65 we have that $\tilde{D}$ is r.c.r. with row and column powers r_i, resp. k_j, given by (70); moreover $\tilde{D}_h = I$.

To illustrate this method, note that the matrix $\bar{D}$, given in (69), is transformed by the e.c.o.'s $\gamma_1 \leftarrow \gamma_1 - s\gamma_2$, $\gamma_1 \leftarrow \gamma_1 + (s-1)\gamma_3$, $\gamma_2 \leftarrow \gamma_2 - \gamma_3$ into

$$\tilde{D}(s) = \begin{bmatrix} s^5 - s & 0 & s \\ s - 1 & s^3 - 1 & 1 \\ 0 & 0 & s^2 \end{bmatrix}$$

where $\tilde{D}$ is now r.c.r. with $\tilde{D}_h = I$ and row and column powers r_i resp. k_j conditioned by

$$r_1 \geq r_2 \geq r_3 \geq 0,\ k_1 \geq k_2 \geq k_3 \geq 0,\ r_1 + k_1 = 5,\ r_2 + k_2 = 3,\ r_3 + k_3 = 2.$$

From the arguments above it is also clear that, by using e.r.o.'s, any r.r. matrix $\bar{D}$ with decreasing row-degrees and $\bar{D}_h = I$ can be converted into a r.c.r. matrix with $\tilde{D}_h = I$ and row and column powers conditioned similarly as in (70).

71 <u>Remark</u>. Criterion (64) suggests another method for making a nonsingular matrix $D \in \mathbb{R}[s]^{\nu\times\nu}$ r.c.r. by e.o.'s over $\mathbb{R}[s]$.

1. Factorize D as

72 $$D = \bar{D}_\ell \, D_- \, \bar{D}_r ,$$

where $\bar{D}_\ell$ and $\bar{D}_r \in \mathbb{R}[s]^{\nu\times\nu}$ and $D_- \in \mathbb{R}(s)^{\nu\times\nu}$ is <u>biproper</u>.

2. Get $\bar{D}_\ell$ r.r. by e.r.o.'s and $\bar{D}_r$ c.r. by e.c.o.'s.

Every nonsingular matrix $D \in \mathbb{R}[s]^{\nu\times\nu}$ has many factorizations (72) since it can be factored as a product of polynomial matrices in many ways (then $D_- = I$ in (72)).

Factorization (72) is useful in the study of feedback systems: the inverse return difference $[I + G_1G_2]^{-1}$ with $G_1 \in \mathbb{R}_p(s)^{m\times n}$ given by a ℓ.f. (D_1, N_1) and $G_2 \in \mathbb{R}_{p,o}(s)^{n\times m}$ given by a r.f. (N_2, D_2) satisfies

73 $$(I + G_1G_2)^{-1} = D_2\, D^{-1}\, D_1,$$

where

74 $$D = D_1D_2 + N_1N_2 = D_1(I + G_1G_2)D_2,$$

with $G_1G_2 \in \mathbb{R}_{p,o}(s)^{m\times m}$. Hence D is of the form (72) and by criterion (64), D is r.c.r. if D_1 is r.r. and D_2 is c.r.

75 <u>Exercise</u>. Show that every nonsingular diagonal matrix $D \in \mathbb{R}[s]^{\nu\times\nu}$ is r.c.r. ∎

The following property is important in that it relates $\partial[\det D]$ and the sum of row and column powers.

76 <u>Fact</u>. A nonsingular matrix $D \in \mathbb{R}[s]^{\nu\times\nu}$ is r.c.r. with row powers $r_i \geq 0$, $i \in \underline{\nu}$, and column powers $k_j \geq 0$, $j \in \underline{\nu}$, if and only if

$$\partial[\det D] = \sum_{i=1}^{\nu} (r_i + k_i).$$ ∎

77 <u>Exercise</u>. Prove Fact 76. (Hint: (a) Argue on the $r_i \geq 0$, $i \in \underline{\nu}$ and the $k_j \geq 0$, $j \in \underline{\nu}$ s.t. $\lim_{s\to\infty} \mathrm{diag}[s^{-r_i}]\, D(s)\, \mathrm{diag}[s^{-k_j}] = D_h \in \mathbb{R}^{\nu\times\nu}$, (b) use $\lim \det \cdots = \det \lim \cdots$.)

We return now to the study of well-formed PMDs.

78 <u>Fact</u>. A nonsingular matrix $D \in \mathbb{R}[s]^{\nu\times\nu}$ which is r.c.r. is such that D^{-1} is proper, or equiv. D has no zeros at ∞.

79 <u>Exercise</u>. Prove Fact 78. (Hint: $D(s)^{-1} = \mathrm{diag}[s^{-k_j}] \cdot \left(\mathrm{diag}[s^{-r_i}]\, D(s)\, \mathrm{diag}[s^{-k_j}]\right)^{-1} \mathrm{diag}[s^{-r_i}]$ with all factors proper.)

80 <u>Fact</u>. Let $D \in \mathbb{R}[s]^{\nu\times\nu}$ be nonsingular and r.c.r. with row powers $r_i \geq 0$, $i \in \underline{\nu}$, and column powers $k_j \geq 0$, $j \in \underline{\nu}$.

81 Let $N_\ell \in \mathbb{R}[s]^{\nu\times n_i}$ s.t. $\partial_{ri}[N_\ell] \leq r_i$ for all $i \in \underline{\nu}$.

82 Let $N_r \in \mathbb{R}[s]^{n_o\times\nu}$ s.t. $\partial_{cj}[N_r] \leq k_j$ for all $j \in \underline{\nu}$.

U.t.c.

(a) $N_rD^{-1} \in \mathbb{R}_p(s)^{n_o\times\nu}$;

(b) $D^{-1}N_\ell \in \mathbb{R}_p(s)^{\nu \times n_i}$;

(c) $N_r D^{-1} N_\ell \in \mathbb{R}_p(s)^{n_o \times n_i}$.

Proof. (a): Observe that

$$N_r(s)\, D(s)^{-1}$$
$$= (N_r(s)\, \mathrm{diag}[s^{-k_j}])\left(\mathrm{diag}[s^{-r_i}]\, D(s)\, \mathrm{diag}[s^{-k_j}]\right)^{-1}(\mathrm{diag}[s^{-r_i}]).$$

Now as $s \to \infty$, all terms on the RHS tend to finite constant matrices because of (82), the fact that D is r.c.r. and (61) and because $r_i \geq 0$ for all $i \in \underline{\nu}$. As a consequence the RHS tends to a finite constant matrix as $s \to \infty$ and so does the LHS. Hence $N_r\, D^{-1} \in \mathbb{R}_p(s)^{n_o \times \nu}$.

(b): Use a similar reasoning as for (a).

(c): Observe that

$$N_r(s)\, D(s)^{-1}\, N_\ell(s)$$
$$= (N_r(s)\, \mathrm{diag}[s^{-k_j}])\left(\mathrm{diag}[s^{-r_i}]\, D(s)\, \mathrm{diag}[s^{-k_j}]\right)^{-1}(\mathrm{diag}[s^{-r_i}]\, N_\ell(s)),$$

where all terms on the RHS tend to finite constant matrices as $s \to \infty$. ∎

We are now able to state a practical corollary to Theorem 52 for obtaining a well-formed PMD.

83 Corollary [Test for a well-formed PMD]. Consider a PMD $\mathcal{D} := [D, N_\ell, N_r, K]$ described by (3.2.1.1) and (3.2.1.2).

U.t.c. if

(a) D is r.c.r. with row powers $r_i \geq 0$, $i \in \underline{\nu}$ and column powers $k_j \geq 0$, $j \in \underline{\nu}$;

(b) $\partial_{ri}[N_\ell] \leq r_i$ for all $i \in \underline{\nu}$,

$\partial_{cj}[N_r] \leq k_j$ for all $j \in \underline{\nu}$;

(c) $K \in \mathbb{R}^{n_o \times n_i}$;

then the PMD $\mathcal{D}$ is well formed. ∎

84 Comments. (a) Notice that once D is r.c.r., the conditions of Corollary 83 can be checked by inspection of the entries of the system-matrix (51). (b) The conditions of Corollary 83 are almost necessary for well-formedness: see

Comments 3.4.45 and Corollary 3.4.46 below.

85 Proof of Corollary 83. We check that condition (55) is satisfied; viz., D^{-1}, N_rD^{-1}, $D^{-1}N_\ell$, and $\hat{H}$ have to be proper. Now condition (a) and Fact 78 imply that D^{-1} is proper. Moreover conditions (a) and (b) and Fact 70 imply that matrices N_rD^{-1}, $D^{-1}N_\ell$, and $N_rD^{-1}N_\ell$ are proper. Now since by condition (c) K is proper, it follows that $\hat{H} = N_rD^{-1}N_\ell + K$ is also proper. ∎

86 Exercise. Show that the PMD $\mathcal{D} = [pI - A, B, C, D]$ satisfies the conditions of Corollary 83. ∎

3.3.2. Exponentially Stable PMDs

We start by considering differential equations.

1 Definition. We say that the differential equation

2 $$D(p)\xi(t) = \theta_\nu \quad t \geq 0,$$

with $D(\cdot) \in \mathbb{R}[p]^{\nu\times\nu}$ nonsingular, is exponentially stable iff for all initial values of $\xi(\cdot)$ and its derivatives at $t = 0-$, (a) $\xi(\cdot)$ does not contain a distribution at $t = 0$ of the form $\sum_{i=0}^{\ell} \xi_i\delta^{(i)}(t)$, where $\xi_i \in \mathbb{R}^\nu$ for all $i \in \underline{\ell}$, and (b) $\xi(\cdot)$ is exponentially decreasing on $\mathbb{R}_+$, or equiv.

3 $\exists\ \alpha > 0$ s.t. $t \mapsto e^{\alpha t}\xi(t)$ is bounded on $\mathbb{R}_+$.

Analysis. By Theorem 2.3.5.38 any solution of (2) which is C^∞ on $(0-, \infty)$ is given by

4 $$\xi(t) = \bar{\Psi}(t)x_\xi(0) \quad \forall t > 0-,$$

where with $n = \partial[\det D]$, $\bar{\Psi}(\cdot)$ is the $\nu \times n$ basis matrix for the solution space and $x_\xi(0) \in \mathbb{R}^n$ is the normalized state of (2) at $t = 0$, obtained by Algorithms 2.3.5.24 and 2.3.5.36, which use the upper triangular Hermite row form of D.

Now, condition (a) of Definition 1 is equivalent to the requirement that differential equation (2) be well formed. Hence for all initial values of $\xi(\cdot)$ and its derivatives at $t = 0-$, $\xi(\cdot)$ is given by (4), $\forall t \geq 0+$ and is at most discontinuous at t = 0 (no distribution at t = 0). Note especially that the state $x_\xi(0) = x_\xi(0+) \in \mathbb{R}^n$ is arbitrary. Therefore, (2) is exponentially stable

if and only if

5 (2) is well formed,

and

6 $\exists\ \alpha > 0$ s.t. $e^{\alpha\cdot}\bar{\Psi}(\cdot)$ is bounded on $\mathbb{R}_+$

It follows now by the construction of the basis matrix $\bar{\Psi}(\cdot)$ in Algorithm 2.3.5.24 that (6) holds if and only if

7 $$Z[\det D] \subset \mathring{\mathbb{C}}_-$$

(note that using the Hermite row form of D the Laplace transform of every element of $\bar{\Psi}(\cdot)$ is strictly proper with poles at the zeros of det D). We know also from Theorem 3.3.1.14 that (5) holds if and only if

8 $$D^{-1} \text{ is proper.}$$

Therefore, we have

9 <u>Theorem</u> [Exponentially stable differential equation]. The differential equation (2) is exponentially stable if and only if

(a)

8 $$D^{-1} \in \mathbb{R}_p(s)^{\nu\times\nu},$$

(b)

7 $$Z[\det D] \subset \mathring{\mathbb{C}}_- .$$ ∎

10 <u>Comment</u>. Note that condition (7)-(8) may be expressed as: <u>the polynomial matrix D</u> (considered as a rational matrix) <u>has no zeros in $\mathbb{C}_+$ and at ∞</u>, or equiv., <u>the rational matrix D^{-1} has no poles in $\mathbb{C}_+$ and at ∞</u>.

We consider now a PMD $\mathcal{D} := [D, N_\ell, N_r, K]$, given by (3.2.1.1)-(3.2.1.2), and recall that in (3.2.1.9) any z-i p-s trajectory $\xi(\cdot)$ of $\mathcal{D}$ is a solution of differential equation (2) which is C^∞ on (0-, ∞) and given by

12 $$\xi(t) = \Psi(t)\, P\, x_\xi(0) \quad \forall t > 0-.$$

Here $\Psi(\cdot)$ is any basis matrix for the C^∞ solution space of (2), $P \in \mathbb{R}^{n\times n}$ is nonsingular, and $x_\xi(0) \in \mathbb{R}^n$ is the normalized state of $\mathcal{D}$ with $n = \partial[\det D]$.

We have then

13 Definition. The PMD $\mathcal{D} := [D, N_\ell, N_r, K]$ given by (3.2.1.1)-(3.2.1.2), is said to be exponentially stable (exp. stable) iff

(a) every z-i p-s trajectory of $\mathcal{D}$ given by (12) is exponentially decreasing on $\mathbb{R}_+$, or equiv.

14 $\forall x_\xi(0) \in \mathbb{R}^n$, $\exists\ \alpha > 0$ s.t.

$$e^{\alpha t}\xi(t) = e^{\alpha t}\Psi(t)P\, x_\xi(0) \text{ is bounded on } \mathbb{R}_+,$$

(b)

15 the PMD $\mathcal{D}$ is well formed.

16 Comment. In Definition 13 condition (14) is the natural requirement for having exponential stability when we study state-space system descriptions, i.e., a PMD $\mathcal{D} := [pI - A, B, C, D]$. Since that PMD is always well formed, no impulsive behavior can occur at $t = 0$, under the conditions of Definitions 3.3.1.25 and 3.3.1.35. For a general PMD $\mathcal{D}$ this is not always true. Therefore, we must require (15): we want a good model.

From the analysis of differential equation (2) and Theorem 3.3.1.52 we have now

18 Theorem [Exponentially stable PMDs]. Consider the PMD $\mathcal{D} := [D, N_\ell, N_r, K]$, given by (3.2.1.1)-(3.2.1.2), with characteristic polynomial det D.

U.t.c.

$\mathcal{D}$ is exp. stable,

if and only if

(a)

$$Z[\det D] \subset \mathring{\mathbb{C}}_-$$

(b)

the rational matrices D^{-1}, $N_r D^{-1}$, $D^{-1}N_\ell$, $\hat{H} = N_r D^{-1} N_\ell + K$ are proper (or equiv. $\mathcal{D}$ is well formed). ∎

An exp. stable PMD has many desirable properties.

19 Theorem. Consider a PMD $\mathcal{D} := [D, N_\ell, N_r, K]$, given by (3.2.1.1)-(3.2.1.2) with transfer function $\hat{H}$.
U.t.c, if $\mathcal{D}$ is exp. stable, then
(a) For every value of $\xi(\cdot)$ and its derivatives at $t = 0-$, the z-i p-s trajectory $\xi(\cdot)$ and the z-i response $y(\cdot)$ are exponentially decreasing on $\mathbb{R}_+$, by which we mean:

$$\exists\ \alpha > 0 \text{ s.t. } t \mapsto e^{\alpha t}\xi(t) \text{ and } t \mapsto e^{\alpha t}y(t) \text{ is bounded on } \mathbb{R}_+.$$

(b) For every value of $\xi(\cdot)$ and its derivatives at $t = 0-$ and for every input $u(\cdot)$ p. suff. diff. on $\mathbb{R}_+$ with $u^{(j)}(0-) = \theta_{n_i}$ for all $j = 0, 1, 2, \cdots$,

20 (i) if $u(\cdot) \in L^\infty$, (equiv. is bounded on $\mathbb{R}_+$), then $\xi(\cdot)$ and $y(\cdot) \in L^\infty$;

21 (ii) if $u(\cdot) \in L^\infty$ and $\lim_{t\to\infty} u(t) = u_\infty \in \mathbb{C}^{n_i}$, then
$\lim_{t\to\infty} \xi(t) = [D^{-1} N_\ell](0)u_\infty$ and $\lim_{t\to\infty} y(t) = \hat{H}(0)u_\infty$;

22 (iii) if $u(\cdot)$ is T-periodic on $\mathbb{R}_+$, then on $\mathbb{R}_+$ there exist T-periodic functions $\xi_p(\cdot)$ and $y_p(\cdot)$ such that, as $t \to \infty$, $\xi(t) \to \xi_p(t)$ and $y(t) \to y_p(t)$; in particular if $u(t) = ke^{j\omega t}$ with $\omega = 2\pi T^{-1}$ and $k \in \mathbb{C}^{n_i}$, then $\xi_p(t) = [D^{-1} N_\ell](j\omega)\, u(t)$ and $y_p(t) = \hat{H}(j\omega)\, u(t)$;

23 (iv) if $u(t) = ke^{zt}$, where $k \in \mathbb{C}^{n_i}$ and $z \in \mathbb{C}_+$, then
$\lim_{t\to\infty} \left\{\xi(t) - [D^{-1} N_\ell](z)\, u(t)\right\} = \theta_\nu$ and $\lim_{t\to\infty} \{y(t) - \hat{H}(z)\, u(t)\} = \theta_{n_o}$. ∎

24 Comment. Note that under the specifications of statement (b) of Theorem 19 the following holds: (a) property (20) guarantees that for a bounded input, the pseudo-state and output will remain bounded; (b) properties (21)-(23) guarantee that asymptotically as $t \to \infty$ the pseudo-state and output of an exp. stable PMD inherit the behavior of the input: convergence, periodicity, exponential behavior. Properties (21)-(23) are used in set point regulation, frequency-response measurement, and tracking.

25 Proof of Theorem 19. Property (a) follows by observing that for every value of $\xi(\cdot)$ and its derivatives at $t = 0-$, the z-i $\hat{\xi}$ and $\hat{y}$ are strictly proper rational functions with poles in $\mathring{\mathbb{C}}_-$.

For the properties under (b) one observes that because of (a), we are done if these properties hold for the z-s p-s trajectory $\xi(\cdot)$ and the z-s response $y(\cdot)$. Now the latter are both given as a convolution of $u(\cdot)$ by an impulse response whose Laplace transform is a proper transfer function with poles in $\mathring{\mathbb{C}}_-$. Hence because of similarity it is sufficient to consider the z-s response $y(\cdot)$ given by

25 $$y(t) = (H * u)(t) = \int_0^t H(\tau)u(t - \tau)\, d\tau,$$

where $H(t) = \mathcal{L}^{-1}[\hat{H}](t)$ and without loss of generality $\hat{H} \in \mathbb{R}_{p,o}(s)^{n_o \times n_i}$ with $P[\hat{H}] \subset \mathring{\mathbb{C}}_-$. Therefore,

26 $$\exists\; k > 0,\ \alpha > 0 \ \text{ s.t. } \ \|H(t)\| \leq ke^{-\alpha t} \quad \forall t \in \mathbb{R}_+.$$

Properties (20)-(23) are now consequences of the following.

(20): With $\|f\|_\infty := \sup\{\|f(t)\| : t \in \mathbb{R}_+\}$, (25) and (26) imply that $\|y\|_\infty \leq k\alpha^{-1}\|u\|_\infty < \infty$.

(21): Defining $u(t) := 0 \ \ \forall t < 0$, we have that

$$y(t) = \int_0^\infty H(\tau)\, u(t - \tau)\, d\tau \quad \forall t \in \mathbb{R}_+,$$

where for all $t \in \mathbb{R}_+$ $\|H(\tau)\, u(t - \tau)\| \leq k\|u\|_\infty e^{-\alpha\tau}$ and $\tau \mapsto k\|u\|_\infty e^{-\alpha\tau}$ is absolutely integrable. Hence by the Lebesgue dominated convergence theorem [Rud.1, Th. 1.34],

$$\lim_{t\to\infty} y(t) = \int_0^\infty H(\tau)[\lim_{t\to\infty} u(t - \tau)]\, d\tau = \hat{H}(0)\, u_\infty.$$

(22): Define, for all $t < 0$, $u(t) := u(t + nT)$, where n is a sufficiently large integer s.t. $t + nT > 0$. As a consequence $u(\cdot)$ is now T-periodic on $\mathbb{R}$ with the same upperbound $\|u\|_\infty$.

Consider now the function $y_p(\cdot)$ on $\mathbb{R}_+$ given by

27 $$y_p(t) := \int_0^\infty H(\tau)\, u(t - \tau)\, d\tau \quad \forall t \in \mathbb{R}_+.$$

Observe that this function is T-periodic and well defined (for the latter

observe that by (26) $\|H(\tau)\ u(t - \tau)\| \leq k\|u\|_\infty e^{-\alpha\tau}$, whence the integral in the RHS of (27) converges absolutely $\forall t \in \mathbb{R}_+$). Now by (25)-(27) for all $t \in \mathbb{R}_+$
$\|y(t) - y_p(t)\| = \|\int_t^\infty H(\tau)\ u(t - \tau)\ d\tau\| \leq k\alpha^{-1}\|u\|_\infty e^{-\alpha t}$.

Hence

$$\lim_{t\to\infty} y(t) = y_p(t).$$

Finally, by the substitution of $u(t) = ke^{j\omega t}$ in (27), we see that

$$y_p(t) = [\int_0^\infty H(\tau)e^{-j\omega\tau}d\tau]\ ke^{j\omega t} = \hat{H}(j\omega)\ u(t) \quad \forall t \in \mathbb{R}_+.$$

(23): Note that with $u(t) = ke^{zt}$ for all $t \in \mathbb{R}_+$ and (25)
$\hat{y}(s) - \hat{H}(z)\ \hat{u}(s) = (s - z)^{-1}[\hat{H}(s) - \hat{H}(z)]\ k$, a strictly proper rational function with poles in $\mathring{\mathbb{C}}_-$. ∎

28 Exercise [Tracking property]. Consider the PMD $\mathcal{D} := [D, N_\ell, N_r, K]$ given by (3.2.1.1) and (3.2.1.2). Let $\mathcal{D}$ be exp. stable.

(a) Show that, for any $\lambda \in \mathbb{C}_+$, $\forall k \in \mathbb{N}$, $\forall u_\alpha \in \mathbb{C}^{n_i}$ with $\alpha = 0, 1, \cdots, k$, for all initial values of $\xi(\cdot)$ and its derivatives at $t = 0-$, the input

$$u(t) := \sum_{\alpha=0}^{k} u_\alpha t^\alpha e^{\lambda t} \qquad t \in \mathbb{R}_+,$$

with $u^{(j)}(0-) = \theta\ \forall j = 0, 1, 2, \cdots$, produces an output $y(\cdot)$ which satisfies as $t \to \infty$

$$\{y(t) - \sum_{\alpha=0}^{k} \hat{H}^{(\alpha)}(\lambda)u_\alpha t^\alpha e^{\lambda t}\} \to \theta$$

exponentially. Note that $\hat{H}$ is the transfer function of $\mathcal{D}$.

(b) Under the conditions of (a) with $k \in \mathbb{N}$ fixed, show that $y(t) - u(t) \to \theta$ exponentially iff the Taylor expansion of $\hat{H}(s)$ about λ reads

$$\hat{H}(s) = I + \frac{(s - \lambda)^{k+1}}{(k + 1)!}\ \hat{H}^{(k+1)}(\lambda) + \cdots;$$

i.e., the transfer function $\hat{H}(s) - I$ must have contact of order k with 0 at λ. ∎

In many applications it is convenient to consider only the transfer function $\hat{H} \in \mathbb{R}(s)^{n_o \times n_i}$ and the I/O-map, (input-output map)

30 $$y(t) = (H * u)(t),$$

where $H(t) = \mathcal{L}^{-1}[\hat{H}(\cdot)]$ is the impulse response, $u(\cdot) : \mathbb{R}_+ \to \mathbb{R}^{n_i}$ is a p. suff. diff. input and $y(\cdot)$ is the output of the system. (When $\hat{H}$ is the transfer function of a PMD, $y(\cdot)$ is the z-s response.)

31 <u>Definition</u>. We say that the transfer function $\hat{H} \in \mathbb{R}(s)^{n_o \times n_i}$ is <u>exponentially stable</u> (exp. stable) iff $\hat{H}$ is <u>proper</u> and $P[\hat{H}] \subset \mathring{\mathbb{C}}_-$, or equiv. $\hat{H} \in R(0)^{n_o \times n_i}$.

32 <u>Comment</u>. Note that $\hat{H}$ is exp. stable iff $\hat{H}$ has no poles in $\mathbb{C}_+$ and at ∞; then the impulse response is exp. decreasing on $t > 0$, with at most a Dirac δ-function at $t = 0$.

The following properties are a consequence of the proof of Theorem 19.

33 <u>Theorem</u>. The I/O map (30) of an exp. stable transfer function $\hat{H}$ has the following properties:

34 (i) if $u(\cdot) \in L^\infty$, then $y(\cdot) \in L^\infty$;

35 (ii) if $u(\cdot) \in L^\infty$ and $\lim_{t\to\infty} u(t) = u_\infty \in \mathbb{R}^{n_i}$ then $\lim_{t\to\infty} y(t) = \hat{H}(0)u_\infty$;

36 (iii) if $u(\cdot)$ is T-periodic on $\mathbb{R}_+$, then there exists a T-periodic function $y_p(\cdot)$ on $\mathbb{R}_+$ s.t. $\lim_{t\to\infty} y(t) = y_p(t)$; in particular for $u(t) = ke^{j\omega t}$ with $k \in \mathbb{C}^{n_i}$ and $\omega := 2\pi T^{-1}$, $y_p(t) = \hat{H}(j\omega)u(t)$;

37 (iv) if $u(t) = ke^{zt}$ for all $t \in \mathbb{R}_+$ with $k \in \mathbb{C}^{n_i}$ and $z \in \mathbb{C}_+$, then $\lim_{t\to\infty} \{y(t) - \hat{H}(z)\, u(t)\} = \theta$. ∎

Theorem 33 shows why exp. stable transfer functions are important. It is, however, most desirable that the properties of $y(\cdot)$ in Theorem 33 also hold for $\xi(\cdot)$ and $y(\cdot)$ under the specifications of Theorem 19; indeed, in many problems $\xi(\cdot)$ has components with physical meaning. Hence the question: When is it that both the transfer function $\hat{H}$ and the underlying PMD $\mathcal{D}$ are exp. stable?

40 Theorem [Exp. stability of transfer function and PMD]. Consider a PMD $\mathcal{D} := [D, N_\ell, N_r, K]$, given by (3.2.1.1)-(3.2.1.2) with transfer function $\hat{H} = N_r D^{-1} N_\ell + K$. Assume that $\hat{H}$ is exp. stable.

U.t.c.

41 $\hat{H}$ and $\mathcal{D}$ are both exp. stable

if and only if

(a)

42 $\mathcal{D}$ is well formed,

or equiv.

43 the rational matrices D^{-1}, $N_r D^{-1}$, $D^{-1}N_\ell$, and $\hat{H}$ are proper;

(b)

44 $\mathcal{D}$ has no unstable hidden modes, by which we mean that every z-i p-s trajectory $\xi(\cdot)$ of $\mathcal{D}$ associated with $\mathbb{C}_+$ eigenvalues of $\mathcal{D}$ must be reachable and observable,

or equiv.

45 $\mathcal{D}$ has no decoupling zeros in $\mathbb{C}_+$.

46 Comment. Note that the exp. stability of both $\hat{H}$ and the PMD is not automatically guaranteed! Note that z-i p-s trajectories associated with $\mathbb{C}_+$-eigenvalues λ of $\mathcal{D}$ are sums of exponential polynomials in t of the form (3.2.1.17), where $\lambda \in \mathbb{C}_+$.

47 Exercise. Consider a PMD $\mathcal{D} := [D, N_\ell, N_r, K]$. Show that $\mathcal{D}$ has no $\mathbb{C}_+$-decoupling zeros iff

48 $$\text{rk}[D(s) \,\vdots\, N_\ell(s)] = \nu \text{ and } \text{rk}\begin{bmatrix} D(s) \\ \hdashline N_r(s) \end{bmatrix} = \nu \quad \forall s \in \mathbb{C}_+.$$

49 Proof of Theorem 40. From Theorem 18 and Definition 31 we note that (41) holds if and only if

50 $$\hat{H} \text{ is exp. stable } \Rightarrow \mathcal{D} \text{ is exp. stable.}$$

Claim 1: We have (42) and (44) $\Leftrightarrow$ (50).

$\Rightarrow$: By assumption $\hat{H}$ is exp. stable, $\mathcal{D}$ is well formed and has no unstable hidden modes. Assume for the purpose of contradiction that $\mathcal{D}$ is not exp. stable; then since $\mathcal{D}$ is well formed, according to Definition 13, $\mathcal{D}$ must have at least one z-i p-s trajectory $\xi(\cdot)$ which is associated with a $\mathbb{C}_+$ eigenvalue λ of $\mathcal{D}$ and by assumption it must be reachable and observable. So by Definitions 3.2.2.3 and 3.2.3.3 there exists an input $u(\cdot)$ with $\hat{u} \in \mathbb{R}[s]^{n_i}$ producing $\xi(\cdot)$ on $t > 0$, with $y(t) = N_r(p)\xi(t) \not\equiv \theta$ on $t > 0$. Note that $\hat{\xi}$ has only poles at $\lambda \in \mathbb{C}_+$ and that the strictly proper part of $\hat{y}$ is nonzero since $y(t) \not\equiv \theta$ on $t > 0$. This implies that $\hat{y}$ must have a pole at $\lambda \in \mathbb{C}_+$; otherwise, since $\hat{y}$ can have only poles at λ, $\hat{y}$ would be polynomial, and this is impossible since $\hat{y}$ has a nonzero strictly proper part. Hence we have that $\exists$ an input $\hat{u} \in \mathbb{R}[s]^{n_i}$ which produces a z-s response $\hat{y} = \hat{H}\hat{u}$ with a pole at $\lambda \in \mathbb{C}_+$. This implies that $\hat{H}$ must have a pole at $\lambda \in \mathbb{C}_+$; (otherwise, $\hat{y}$ cannot have a pole at $\lambda \in \mathbb{C}_+$). Since an exp. stable $\hat{H}$ has no $\mathbb{C}_+$ poles, we have a contradiction. Hence $\mathcal{D}$ must be exp. stable. ∎

$\Leftarrow$: By assumption (50) holds. Then (42) must hold by Definition 13. Note also that, since $\mathcal{D}$ is exp. stable, by Theorem 18, $Z[\det D] \subset \mathring{\mathbb{C}}_-$: it follows that $\mathcal{D}$ has no $\mathbb{C}_+$ eigenvalues. Therefore, (44) is trivially satisfied since there are no z-i p-s trajectories associated with $\mathbb{C}_+$ eigenvalues. ∎

Since Claim 1 holds now and (42) $\Leftrightarrow$ (43) by Theorem 3.3.1.52, we are done if we show (44) $\Leftrightarrow$ (45).

Claim 2: We have that (44) $\Leftrightarrow$ (45).

A rigorous bookkeeping of linearly independent z-i p-s trajectories $\xi(\cdot)$ of $\mathcal{D}$ associated with an eigenvalue $\lambda \in \mathbb{C}$ uses the Jordan chain relation [Goh.1], given by

51 $$\sum_{i=0}^{\ell} \frac{1}{(\ell - i)!} D^{(\ell-i)}(\lambda)\xi_i = \theta_\nu \quad \text{for } \ell = 0, 1, 2, \cdots.$$

This produces generalized eigenvectors $\xi_\ell \in \mathbb{C}^\nu$ for $\ell = 0, 1, 2, \cdots$ and generates linearly independent z-i p-s trajectories associated with $\lambda \in \mathbb{C}$ of the form

52 $$\xi_\ell(t) = \sum_{i=0}^{\ell} \xi_i \frac{t^{\ell-i}}{(\ell - i)!} e^{\lambda t} \quad \text{for } \ell = 0, 1, 2, \cdots.$$

We assume here for reasons of simplicity that the n eigenvalues λ_i of $\mathcal{D}$ are distinct. Under this assumption there exists a bijection between the n eigenvalues $\lambda_i \in \mathbb{C}$ of $\mathcal{D}$ and n eigenvectors of fixed direction $\xi_i \in \mathbb{C}^{\nu}$ satisfying

53 $$D(\lambda_i)\xi_i = \theta \quad \xi_i \neq \theta.$$

In this manner a basis for the space of z-i p-s trajectories $\xi(\cdot)$ is given by

54 $$\xi_i(t) = \xi_i e^{\lambda_i t} \qquad i \in \underline{n}.$$

Let now L be a g.c.ℓ.d. of (D, N_ℓ) and R a g.c.r.d. of (N_r, D), whence D can be factorized as

55 $$D = L\bar{D} \quad \text{and} \quad D = \tilde{D}R,$$

where by assumption D and all factors have distinct eigenvalues (i.e., the determinants have distinct roots). Note also by Definitions 3.2.2.33 and 3.2.3.22 that i-d zeros and o-d zeros are eigenvalues of $\mathcal{D}$ that are eigenvalues of L and R, resp. Furthermore, by Theorems 3.2.2.10 and 3.2.3.10 we have that (44) is equivalent to the fact that

56 $$\forall \lambda_i \in \mathbb{C}_+ \quad \bar{D}(p)\xi_i(t) = \theta \quad \forall t \geq 0 \text{ and } R(p)\xi_i(t) \not\equiv \theta \text{ on } t > 0,$$

where $\xi_i(\cdot)$, $\bar{D}$ and R are given by (53)-(55). Using this information and the assumption of distinct eigenvalues, it follows that (56) is equivalent to

$$\forall \lambda_i \in \mathbb{C}_+ \ \det L(\lambda_i) \neq 0 \text{ and } \det R(\lambda_i) \neq 0.$$

Hence we have shown that (44) $\Leftrightarrow$ (45), and Claim 2 is proved.

End of Proof

57 Important Remark. In many applications we will encounter exponentially stable transfer functions. It is of crucial importance that the underlying PMD also be exp. stable. Consequently, we shall implicitly assume that the conditions of Theorem 40 are satisfied or we will check them by analysis (see, e.g., the plant assumption in (8.2.27); see also Fact 4.2.77 and Theorem 4.2.84).

3.4. Transfer Functions: Right-Left Fractions; Internally Proper Fractions

The objective of this section is to describe transfer functions which are not necessarily right or left fractions and similar to $C(sI - A)^{-1}B$ as obtained in state space models.

1 Definitions. Let $H \in \mathbb{R}(s)^{n_o \times n_i}$. We say that the triple of polynomial matrices $(N_r, D, N_\ell) \in E(\mathbb{R}[s])$ is right-left-coprime fraction (r.ℓ.c.f.) of H iff

(a) $\det D(s) \not\equiv 0$,

(b) $H = N_r D^{-1} N_\ell$,

(c) (N_r, D) is r.c. and (D, N_ℓ) is ℓ.c.

If we do not require (c), we say that (N_r, D, N_ℓ) is a right-left fraction (r.ℓ.f.) of H.

2 Comment. Note that r.ℓ.f.'s are naturally associated with a PMD $[D, N_\ell, N_r, 0]$, while a r.ℓ.c.f. is associated with such a minimal PMD. This association with PMDs allows to state the following result, whose proof is contained in that of Theorem 3.2.4.6.

3 Theorem [Poles of a r.ℓ.f.]. Let $H \in \mathbb{R}(s)^{n_o \times n_i}$ have a r.ℓ.f. (N_r, D, N_ℓ).

U.t.c.

(a)

4 $$p \in P[H] \Rightarrow \det D(p) = 0.$$

(b) If (N_r, D, N_ℓ) is a r.ℓ.c.f., then

5 $$p \in P[H] \Leftrightarrow \det D(p) = 0.$$ ∎

In view of Theorem 3 and Definition 3.3.2.31, we also have

8 Theorem [Exp. stable r.ℓ.c.f.s]. Let $H \in \mathbb{R}_p(s)^{n_o \times n_i}$ have a r.ℓ.c.f. (N_r, D, N_ℓ); then

$$H \text{ is exp. stable}$$

$$\Leftrightarrow$$

$$Z[\det D] \subset \mathring{\mathbb{C}}_-.$$ ∎

10 Comment. A PMD $\mathcal{D} := [D, N_\ell, N_r, 0]$ gives rise to a transfer function $H = N_r D^{-1} N_\ell$ which is a r.ℓ.f. If the PMD is minimal, then this r.ℓ.f. has to be coprime. If the PMD is well formed, then according to Theorem 3.3.1.52, it is not sufficient for H to be proper. Hence the following definition, which makes the association of a r.ℓ.f. with a well-formed PMD automatic.

11 Definition. We say that the r.ℓ.f. (N_r, D, N_ℓ) of $H \in \mathbb{R}(s)^{n_o \times n_i}$ is internally proper (int. pr.) iff D^{-1}, $N_r D^{-1}$, $D^{-1} N_\ell$, and $H = N_r D^{-1} N_\ell$ are proper rational matrices.

12 Fact. (a) A ℓ.f. (D_ℓ, N_ℓ) of $H \in \mathbb{R}(s)^{n_o \times n_i}$ is int. pr. iff D^{-1} and $H = D^{-1} N_\ell$ are proper rational matrices.
(b) A r.f. (N_r, D_r) of $H \in \mathbb{R}(s)^{n_o \times n_i}$ is int. pr. iff D_r^{-1} and $H = N_r D_r^{-1}$ are proper rational matrices.

Proof. A ℓ.f. (D_ℓ, N_ℓ) is r.ℓ.f. (I, D_ℓ, N_ℓ) and a r.f. (N_r, D_r) is a r.ℓ.f. (N_r, D_r, I).

We shall now characterize int. pr. right-left-fractions. For this we need the following definition and preliminary results.

14 Definition. Let $H \in \mathbb{R}(s)^{n_o \times n_i}$. We say that the triple of polynomial matrices $(D_\ell, N, D_r) \in E(\mathbb{R}[s])$ is a left-right-coprime fraction, (ℓ.r.c.f.), of H, iff

(a) $\det D_\ell(s) \not\equiv 0$, $\det D_r(s) \not\equiv 0$,

(b) $H = D_\ell^{-1} N D_r^{-1}$,

(c) (D_ℓ, N) is ℓ.c. and (N, D_r) is r.c.

If we do not require (c), we say that (D_ℓ, N, D_r) is a left-right fraction (ℓ.r.f.) of H.

15 Fact [zeros of ℓ.r.c.f.'s]. Let (D_ℓ, N, D_r) be a ℓ.r.c.f. of $H \in \mathbb{R}(s)^{n \times n}$; then $z \in \mathbb{C}$ is a zero of H iff $\det N(z) = 0$.

Proof. $H = D_\ell^{-1} N D_r^{-1}$ has a zero at $z \in \mathbb{C}$ iff $H^{-1} = D_r N^{-1} D_\ell$ has a pole at $z \in \mathbb{C}$. Now (D_r, N, D_ℓ) is r.ℓ.c.f. of H^{-1}. Hence the fact follows by Theorem 3. ∎

16 Theorem [D^{-1} proper]. Let $D \in \mathbb{R}[s]^{\nu\times\nu}$ be nonsingular.
U.t.c.

(a) $$D^{-1} \in \mathbb{R}_p(s)^{\nu\times\nu}$$

iff

17 $$D = D_{cr}\, D_-\, D_{rr},$$

where $D_{cr} \in \mathbb{R}[s]^{\nu\times\nu}$ is column-reduced, $D_- \in \mathbb{R}(s)^{\nu\times\nu}$ is biproper, and $D_{rr} \in \mathbb{R}[s]^{\nu\times\nu}$ is row-reduced.

(b) $$D^{-1} \in \mathbb{R}_p(s)^{\nu\times\nu}$$

iff

$\exists$ unimodular matrices L and R $\in E(\mathbb{R}[s])$, obtained by e.o.'s over $\mathbb{R}[s]$, s.t.

18 (i) LDR is r.c.r. with row powers r_i, $i \in \underline{\nu}$, and column powers k_j, $j \in \underline{\nu}$.
(ii) $\partial_{ri}[L] \le r_i$ $\forall i \in \underline{\nu}$ and $\partial_{cj}[R] \le k_j$ $\forall j \in \underline{\nu}$.

19 Comment. Characterization (b) shows that a r.c.r. matrix is a standard form for a matrix $D \in \mathbb{R}[s]^{\nu\times\nu}$ s.t. D^{-1} is proper, modulo "economical" e.o.'s.

21 Proof of Theorem 16. (a) Setting $s = \lambda^{-1} \in \mathbb{C}$, we have that

22 $D(s)^{-1}$ is proper iff $D(\lambda^{-1}) \in \mathbb{R}_p(s)^{\nu\times\nu}$ has no zero at $\lambda = 0$.
Now a ℓ.r.c.f. of $D(\lambda^{-1})$ is of the form

23 $$D(\lambda^{-1}) = D_\ell(\lambda)^{-1}\check{D}(\lambda)D_r(\lambda)^{-1},$$

where all matrices on the RHS are polynomial in λ. Moreover, D_ℓ and D_r have Smith forms

24 $$D_\ell(\lambda) = L_\ell(\lambda)\ \mathrm{diag}[\lambda^{\gamma_j}]R_\ell(\lambda),$$

25 $$D_r(\lambda) = L_r(\lambda)\ \mathrm{diag}[\lambda^{\rho_i}]R_r(\lambda),$$

where L_r, L_ℓ, R_r, and R_ℓ are unimodular (note that $D(\lambda^{-1})$ has only poles at $\lambda = 0$). Hence by (22) and Fact 15,

26 $$D(s)^{-1} \text{ is proper iff } \det \check{D}(0) \neq 0,$$

or setting $s = \lambda^{-1}$

27 $D(s)^{-1}$ is proper iff $\check{D}(s^{-1})$ is biproper.

Notice that by setting $s = \lambda^{-1}$ in (24)-(25)

28 $D_\ell(s^{-1})$ and $D_r(s^{-1})$ are proper.

Claim 1. We have that

29 $$D_\ell(s^{-1}) = \tilde{L}(s)\, D_{cr}(s)^{-1} \text{ and } D_r(s^{-1}) = D_{rr}(s)^{-1}\tilde{R}(s)$$

where matrices L and R are biproper and matrices D_{cr} and D_{rr} are polynomial matrices which are column-reduced and row-reduced resp. Because of the symmetry in formulas (24)-(25) we shall only prove (29) for D_ℓ.

For this purpose, set $s = \lambda^{-1}$ in (24): we have

30 $$D_\ell(s^{-1}) = L_\ell(s^{-1})\left\{R_\ell(s^{-1})^{-1}\mathrm{diag}[s^{\gamma_i}]\right\}^{-1},$$

where matrices $L_\ell(s^{-1})$ and $R_\ell(s^{-1})^{-1}$ are biproper. It follows now that

31 $$R_\ell(s^{-1})^{-1}\, \mathrm{diag}[s^{\gamma_i}] = D_{csp}(s) + D_{cr}(s),$$

where $D_{csp} \in \mathbb{R}(s)^{\nu\times\nu}$ is strictly proper and $D_{cr} \in \mathbb{R}[s]^{\nu\times\nu}$ is c.r. Hence by (30) and (31)

$$D_\ell(s^{-1}) = L_\ell(s^{-1})[I + D_{cr}(s)^{-1}D_{csp}(s)]^{-1}D_{cr}(s)^{-1}$$
$$=: \tilde{L}(s)\, D_{cr}(s)^{-1},$$

where $\tilde{L}$ is biproper. Hence (29) is proved for D_ℓ and by the noted symmetry in (24)-(25) Claim 1 follows. ∎

Now using (27), (23), and (29) with $s = \lambda^{-1}$, it follows by

$$D_{cr}(s)^{-1}\, D(s)\, D_{rr}(s)^{-1} = \tilde{L}(s)^{-1}\, \check{D}(s^{-1})\, \tilde{R}(s)^{-1},$$

where $\tilde{L}$ and $\tilde{R}$ are biproper, that

$$D(s)^{-1} \text{ is proper iff } D_- := D_{cr}^{-1}\, D\, D_{rr}^{-1} \text{ is biproper.}$$

This establishes criterion (17).
(b): Notice in criterion (17) that D_{cr}^{-1} and D_{rr}^{-1} are proper.
Hence $\exists$ unimodular matrices L and R s.t.

32 $LD_{cr} = \bar{D}_{rr}$ and $D_{rr}R = \bar{D}_{cr}$

with

$\bar{D}_{rr}$ r.r. with row degrees r_i,

$\bar{D}_{cr}$ c.r. with column degrees k_j,

33 $\partial_{ri}[L] \leq r_i \quad \forall i \in \underline{\nu}$,

34 $\partial_{cj}[R] \leq k_j \quad \forall j \in \underline{\nu}$.

As a consequence by (17) and (32), $LDR = \bar{D}_{rr}\, D_-\, \bar{D}_{cr}$; s.t. by criterion (3.3.1.64)

35 LDR is r.c.r. with row powers r_i and column powers k_j.

Hence by (33)-(35) the necessity of criterion (18) follows. For sufficiency note that by criterion (3.3.1.63)

$$D^{-1} = \{R(\text{diag}[s^{k_j}])^{-1}\}\, D_-^{-1}\{(\text{diag}[s^{r_i}])^{-1}L\},$$

where all factors on the RHS are proper.

End of Proof

We are now ready to characterize int. pr. r.ℓ.f.'s. In all results below we stress the relation with well-formed PMDs (see Definition 11).

40 <u>Theorem</u> [Int. proper r.ℓ.f.'s]. Let $H \in \mathbb{R}(s)^{n_o \times n_i}$ have a r.ℓ.f. (N_r, D, N_ℓ), where $D \in \mathbb{R}[s]^{\nu\times\nu}$.

U.t.c.

41 the r.ℓ.f. (N_r, D, N_ℓ) of H is <u>internally proper</u> or equiv.

42 the PMD $\mathcal{D} = [D, N_\ell, N_r, 0]$ is <u>well formed</u>,

(a) if and only if

43 $$D = D_{cr} D_- D_{rr},$$

where $D_- \in \mathbb{R}(s)^{\nu\times\nu}$ is biproper and D_{cr} and D_{rr} are polynomial matrices, with D_{cr} column-reduced and D_{rr} row-reduced, s.t.

$$N_r D_{rr}^{-1} \text{ and } D_{cr}^{-1} N_\ell \text{ are proper rational matrices;}$$

(b) or equivalently,
if and only if
$\exists$ unimodular matrices L and $R \in E(\mathbb{R}[s])$ s.t.

44 (i) LDR is r.c.r. with row powers r_i, $i \in \underline{\nu}$, and column powers k_j, $j \in \underline{\nu}$,

(ii) $\partial_{ri}[L] \leq r_i \ \ \forall i \in \underline{\nu}$ and $\partial_{cj}[R] \leq k_j \ \forall j \in \underline{\nu}$,

(iii) $\partial_{ri}[LN_\ell] \leq r_i \ \forall i \in \underline{\nu}$ and $\partial_{cj}[N_r R] \leq k_j \ \forall j \in \underline{\nu}$.

45 <u>Comments</u>. (a) Notice that criterion (17) resp. (18) is contained in criterion (43) resp. (44): this is caused by the fact that by Definition 11 D^{-1} has to be proper: the additional conditions are present to ensure that $N_r D^{-1}$, $D^{-1}N_\ell$ and $\hat{H} = N_r D^{-1} N_\ell$ are proper.
(b) Criterion (44) shows that the conditions of Corollary 3.3.1.83 for guaranteeing that a PMD $\mathcal{D} = [D, N_\ell, N_r, K]$ with K constant were almost necessary: what is needed are appropriate e.o.'s on N_r, D, N_ℓ. Indeed, we have

46 <u>Corollary</u>. [Well-formed PMD's with K constant]. Consider the PMD $\mathcal{D} = [D, N_\ell, N_r, K]$, given by (3.2.1.1)-(3.2.1.2), where $K \in \mathbb{R}^{n_o\times n_i}$
U.t.c. $\mathcal{D}$ is well formed iff criterion (43) or criterion (44) holds. ∎

50 <u>Proof of Theorem 40</u>. According to Theorem 3.3.1.52 and Definition 11, (41) and (42) are equivalent.
(a): Criterion (43) is established as follows:
We have:

41 the r.ℓ.f. (N_r, D, N_ℓ) of H is internally proper
iff
with $s = \lambda^{-1}$, the square proper rational matrix $\tilde{D}(\lambda^{-1})$ given by

51 $$\tilde{D}(\lambda^{-1}) = \begin{bmatrix} I & -N_r(\lambda^{-1}) & 0 \\ 0 & D(\lambda^{-1}) & N_\ell(\lambda^{-1}) \\ 0 & 0 & I \end{bmatrix}$$

has no zero at $\lambda = 0$.

Now there exists a ℓ.r.c.f. of $\tilde{D}(\lambda^{-1})$ of the form

52 $$\tilde{D}(\lambda^{-1}) = \begin{bmatrix} I & 0 & 0 \\ 0 & D_\ell(\lambda) & 0 \\ 0 & 0 & I \end{bmatrix}^{-1} \begin{bmatrix} I & -\check{N}_r(\lambda) & 0 \\ 0 & \check{D}(\lambda) & \check{N}_\ell(\lambda) \\ 0 & 0 & I \end{bmatrix} \begin{bmatrix} I & 0 & 0 \\ 0 & D_r(\lambda) & 0 \\ 0 & 0 & I \end{bmatrix}^{-1},$$

where all matrices on the RHS of (52) are polynomial and $D_\ell(\lambda)$ and $D_r(\lambda)$ have Smith forms (24)-(25). By Fact 15 and setting $s = \lambda^{-1}$, we therefore have by (51)-(52) that

53 $$\begin{cases} \text{(41) holds} \\ \text{iff} \\ \check{D}(s^{-1}) \text{ is biproper and } \check{N}_r(s^{-1}) \text{ and } \check{N}_\ell(s^{-1}) \text{ are proper.} \end{cases}$$

Using Claim 1 of the Proof of Theorem 16 and (51)-(53), it is then possible to establish criterion (43).

(b): Criterion (44) is established using criterion (43) where D_{rr}^{-1} and D_{cr}^{-1} are proper. Hence there exist unimodular matrices L and R, obtained by e.o.'s, s.t. (32) holds with all its properties as listed in the proof of Theorem 16. The remainder of the proof for establishing criterion (44) is left as an exercise. ∎

We conclude this section by two corollaries characterizing int. pr. left- or right-fractions. Their proof is left as an exercise.

56.L <u>Corollary</u> [Int. proper left fractions]. Let $H \in \mathbb{R}(s)^{n_o \times n_i}$ have a ℓ.f. (D_ℓ, N_ℓ).

U.t.c.

The ℓ.f. (D_ℓ, N_ℓ) of H is <u>int. pr.</u>,

or equiv.

the PMD $\mathcal{D} = [D_\ell, N_\ell, I, 0]$ is well formed

(a) if and only if

$$D_\ell = D_{cr} D_- ;$$

57.L where $D_- \in \mathbb{R}(s)^{\nu\times\nu}$ is biproper and $D_{cr} \in \mathbb{R}[s]^{\nu\times\nu}$ is column-reduced s.t.

$$D_{cr}^{-1} N_\ell \text{ is proper,}$$

(b) or equivalently

if and only if

$\exists$ a unimodular matrix L, obtained by e.r.o.'s, s.t.

58.L
(i) $L\,D_\ell$ is row-reduced with row degrees r_i,

(ii) $\partial_{ri}[L] \le r_i \ \forall i \in \underline{n}_o$ and $\partial_{ri}[LN_\ell] \le r_i \ \forall i \in \underline{n}_o$. ∎

56.R Corollary [Int. proper right fractions]. Let $H \in \mathbb{R}(s)^{n_o\times n_i}$ have a r.f. (N_r, D_r).
U.t.c.

The r.f. (N_r, D_r) of H is int. pr.,

or equiv.

the PMD $\mathcal{D} = [D_r, I, N_r, 0]$ is well formed,

(a)

if and only if

$$D_r = D_- D_{rr},$$

57.R where $D_- \in \mathbb{R}(s)^{\nu\times\nu}$ is biproper and $D_{rr} \in \mathbb{R}[s]^{\nu\times\nu}$ is row-reduced s.t.

$$N_r D_{rr}^{-1} \text{ is proper}$$

(b) or equivalently

if and only if

$\exists$ a unimodular matrix R, obtained by e.c.o.'s, s.t.

58.R
(i) $D_r R$ is column-reduced with column degrees k_j,

(ii) $\partial_{cj}[R] \le k_j \ \forall j \in \underline{n}_i$ and $\partial_{cj}[N_r R] \le k_j \ \forall j \in \underline{n}_i$. ∎

59 Exercise. Prove Corollaries 56.L and 56.R. (Hint: For Corollary 56.L observe that H has an int. pr. ℓ.f. (D_ℓ, N_ℓ) iff $\begin{bmatrix} D_\ell(\lambda^{-1}) & N_\ell(\lambda^{-1}) \\ 0 & I \end{bmatrix}$ has no zero at $\lambda = 0$; moreover, this matrix admits a ℓ.c.f. of the form

$$\begin{bmatrix} \tilde{D}_\ell(\lambda) & 0 \\ 0 & I \end{bmatrix}^{-1} \begin{bmatrix} \check{D}_\ell(\lambda) & \check{N}_\ell(\lambda) \\ 0 & I \end{bmatrix} \cdots).$$

Chapter 4. Interconnected Systems

4.1. Introduction

This chapter contains a systematic development of the exponential stability of an interconnection of subsystems; it is then applied to feedback systems.

Section 2 describes an interconnected system Σ consisting of a number of subsystems. The assumptions on interconnection structure and well posedness lead to the description of a feedback system in terms of transfer functions. The assumption on an underlying PMD for each subsystem leads to the description of an overall well-formed PMD with given closed-loop characteristic polynomial (4.2.69). This leads to the exponential stability of Σ in terms of the exp. stability of a PMD, (Th. 4.2.73). Finally exponential stability in terms of transfer functions is discussed (Th. 4.2.84).

Section 3 treats in detail feedback system stability. Stability Theorem 4.3.6 specifies 4 closed-loop characteristic polynomials. Corollary 4.3.26 considers the case when the compensator is given by a fraction which is not necessarily coprime.

Section 4 establishes three important properties of feedback systems: (a) the determinant of the return-difference matrix is the ratio of the closed-loop over the open-loop characteristic polynomial; (b) plant-$\mathbb{C}_+$-zeros limit the amount of loop gain a feedback system can tolerate without becoming unstable; and (c) feedback cannot remove the plant-$\mathbb{C}_+$-zeros.

4.2. Exponential Stability of an Interconnection of Subsystems

Consider an <u>interconnected system Σ</u> of μ given subsystems described by their transfer function matrices

$$G_k \in \mathbb{R}_p(s)^{n_{ok} \times n_{ik}} \qquad k = 1, 2, \cdots, \mu.$$

For example, consider the interconnection shown in Fig. 1.

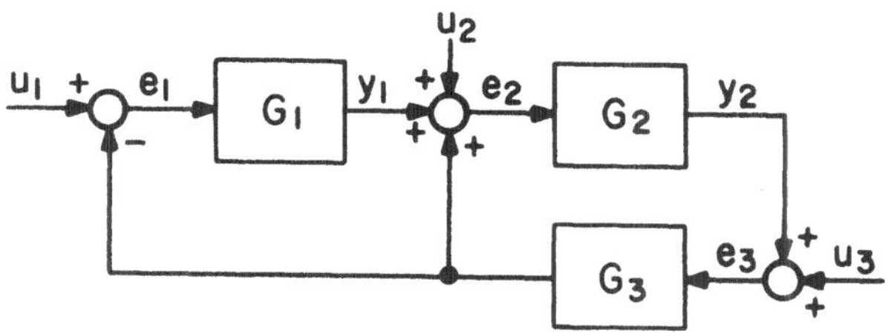

Fig. 1. An interconnected system Σ.

2 Convention. In this and the following sections we often work in the frequency domain s.t. all quantities are Laplace transforms of impulse responses and time functions on $\mathbb{R}_+$. We shall therefore omit the superscript $\hat{\ }$.

We describe now the interconnected system Σ.

3 Assumption IS. For each subsystem $G_k \in \mathbb{R}_p(s)^{n_{ok} \times n_{ik}}$, $k \in \underline{\mu}$, we have the interconnection structure shown in Fig. 2: to each subsystem G_k with input e_k and output y_k we associate a summing node with the following characteristics:
(S1) its output is the subsystem input e_k;

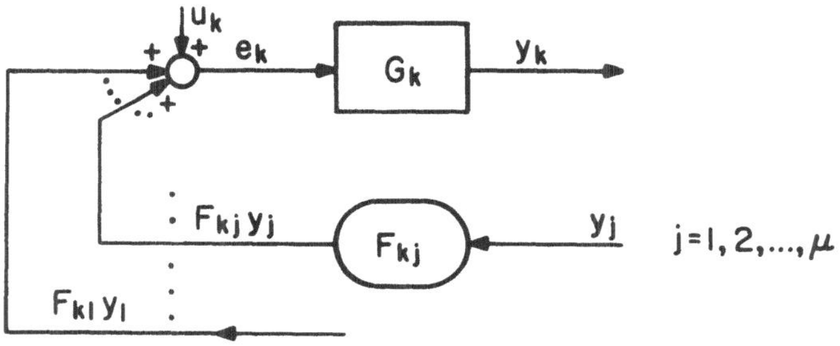

Fig. 2. Associated interconnection structure for subsystem G_k.

(S2) its inputs are:

(a) an exogenous input u_k (always assigned);

(b) other inputs which are feedbacks of the form

4 $$F_{kj}y_j \quad \text{for } j \in \underline{\mu},$$

where $F_{kj} \in \mathbb{R}^{n_{ik} \times n_{oj}}$ denotes the constant gain matrix from y_j to the kth summing node (some of these gain matrices may be zero). ∎

5 Implications of Assumption IS. The subsystems $G_k \in \mathbb{R}_p(s)^{n_{ok} \times n_{ik}}$, $k \in \underline{\mu}$, are interconnected according to the equations

6 $$e_k = u_k + \sum_{j=1}^{\mu} F_{kj}y_j,$$

$$\text{for } k \in \underline{\mu}.$$

7 $$y_k = G_k e_k$$

Hence by aggregation, i.e., by defining global quantities:

8 $$n_i := \sum_{k=1}^{\mu} n_{ik}, \quad n_o := \sum_{k=1}^{\mu} n_{ok},$$

9 $$u := [u_1^T \,\vdots\, \cdots \,\vdots\, u_\mu^T]^T \text{ having dimension } n_i,$$

10 $$e := [e_1^T \,\vdots\, \cdots \,\vdots\, e_\mu^T]^T \text{ having dimension } n_i,$$

11 $$y := [y_1^T \,\vdots\, \cdots \,\vdots\, y_\mu^T]^T \text{ having dimension } n_o,$$

12 $$F := [F_{kj}]_{k,j\in\underline{\mu}} \in \mathbb{R}^{n_i \times n_o},$$

13 $$G := \text{block diag}[G_k]_{k=1}^{\mu} \in \mathbb{R}_p(s)^{n_o \times n_i}.$$

Then, description (6)-(7) of interconnected system Σ is transformed into an equivalent description:

14 $$e = u + Fy,$$

15 $$y = Ge,$$

which describes the feedback system shown in Fig. 3.

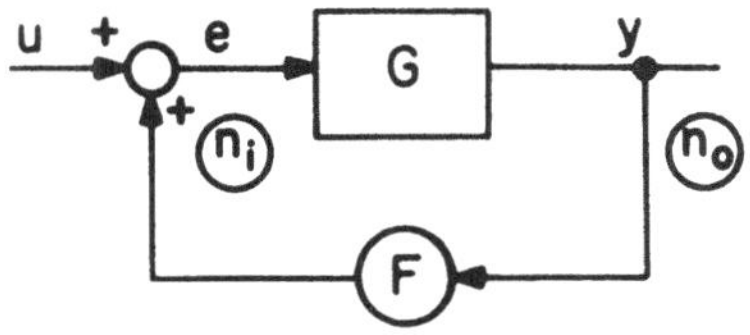

Fig. 3. Interconnected system Σ: the feedback system obtained after aggregation.

For this reason we shall call (i) u, e, y, defined by (8)-(11), the input, error, and output resp. of Σ; (ii) F, defined by (4), (8), and (12), the (feedback) gain matrix of Σ and (iii) G, defined by (1), (8), and (13), the open-loop transfer function of Σ.

For the example of Fig. 1:

$$G = \begin{bmatrix} G_1 & 0 & 0 \\ 0 & G_2 & 0 \\ 0 & 0 & G_3 \end{bmatrix} \begin{matrix} n_{o1} \\ n_{o2} \\ n_{o3} \end{matrix}, \quad (\text{columns } n_{i1}, n_{i2}, n_{i3}) \qquad F = \begin{bmatrix} 0 & 0 & -I \\ I & 0 & -I \\ 0 & I & 0 \end{bmatrix} \begin{matrix} n_{i1} \\ n_{i2} \\ n_{i3} \end{matrix} \quad (\text{columns } n_{o1}, n_{o2}, n_{o3}).$$

∎

Equations (14)-(15) and Fig. 3 show that the interconnected system Σ has the input-error and input-output transfer functions

16 $$H_{eu} : u \mapsto e, \qquad H_{yu} : u \mapsto y, \text{ resp.};$$

where

17 $$H_{eu} = (I - FG)^{-1} \in \mathbb{R}_p(s)^{n_i \times n_i}; \; H_{yu} = G(I - FG)^{-1} \in \mathbb{R}_p(s)^{n_o \times n_i}.$$

Note that

18 $$H_{eu} = I + FH_{yu}.$$

We note in Fig. 3 that all closed-loop transfer functions of Σ, i.e., from u into e, y, Fy resp. are given by H_{eu}, H_{yu}, $FH_{yu} = H_{eu} - I$.

We now introduce a well-posedness assumption.

19 Assumption WP. With F and G defined by (1), (4), (8), and (12)-(13) we assume that interconnected system Σ is s.t.

20 $$\det(I - FG)(\infty) \neq 0.$$

21 Comment. Assumption WP is a necessary and sufficient condition for all closed-loop transfer functions H_{yu}, H_{eu}, and FH_{yu} to be proper. ∎

For stability purposes, in addition to Assumptions IS and WP, it is also desirable to have an assumption which permits us to represent the subsystem transfer matrices G_k as polynomial matrix fractions which are generated by subsystem PMD's $\mathcal{D}_k$ such that G_k is exp. stable if and only if $\mathcal{D}_k$ is exp. stable (see Chapter 3). We have, therefore,

25 Assumption PMD. For k = 1, 2, ⋯, μ each subsystem transfer function $G_k \in \mathbb{R}_p(s)^{n_{ok} \times n_{ik}}$ has a right-left fraction $(N_{rk}, D_k, N_{\ell k})$ generated by a PMD $\mathcal{D}_k := [D_k, N_{\ell k}, N_{rk}, 0]$ which is well formed and has no unstable hidden modes. $\forall k \in \underline{\mu}$, PMD $\mathcal{D}_k$ is described by the time-domain equations

26 $$D_k(p)\xi_k(t) = N_{\ell k}(p)e_k(t),$$

$$t \geq 0,$$

27 $$y_k(t) = N_{rk}(p)\xi_k(t)$$

where $\xi_k(\cdot)$, of dimension ν_k, is the pseudo-state of $\mathcal{D}_k$, whence in the frequency domain,

(a)

28 $$\begin{cases} (N_{rk}, D_k, N_{\ell k}) \in \mathbb{R}[s]^{n_{ok} \times \nu_k} \times \mathbb{R}[s]^{\nu_k \times \nu_k} \times \mathbb{R}[s]^{\nu_k \times n_i} \\ \text{with} \\ \quad \det D_k(s) \neq 0, \end{cases}$$

(b)

29 $$G_k = N_{rk} D_k^{-1} N_{\ell k}.$$ ∎

32 Fact. Assumption PMD is satisfied if and only if $\forall k = 1, 2, \cdots, \mu$, the subsystem transfer function $G_k \in \mathbb{R}(s)^{n_{ok} \times n_{ik}}$ has a r.ℓ.f. $(N_{rk}, D_k, N_{\ell k})$ (28)-(29), s.t.

33 (i) $(N_{rk}, D_k, N_{\ell k})$ is an internally proper r.ℓ.f. of G_k,

34 (ii) $\text{rk}[D_k \,\vdots\, N_{\ell k}](s) = \nu_k \quad \forall s \in \mathbb{C}_+$,

35 (iii) $\text{rk}\begin{bmatrix} D_k \\ \hdashline N_{rk} \end{bmatrix}(s) = \nu_k \quad \forall s \in \mathbb{C}_+$. ■

Proof. According to Theorem 3.3.1.52 and Definition 3.4.11, condition (i) is necessary and sufficient for the PMD $\mathcal{D}_k$ (26)-(27) to be well formed. According to Exercise 3.3.2.47, conditions (ii) and (iii) are necessary and sufficient for the PMD $\mathcal{D}_k$ to have no $\mathbb{C}_+$-decoupling zeros. ■

36 Comment. Assumption PMD allows a left- or a right-coprime fraction for each G_k. Notice that in that case the conditions for having $\mathcal{D}_k$ well formed are simple (see Corollaries 3.4.56.L and 3.4.56.R). ■

We shall now study the properties of an interconnected system Σ under the three Assumptions (3), (19), and (25). First, let us note that by aggregation, i.e., defining

38 $$\nu = \sum_{k=1}^{\mu} \nu_k,$$

39 $$\xi := [\xi_1^T \,\vdots\, \cdots \,\vdots\, \xi_\mu^T]^T \text{ having dimension } \nu,$$

40 $$D := \text{block diag}[D_1, \cdots, D_\mu] \in \mathbb{R}[s]^{\nu \times \nu},$$

41 $$N_\ell := \text{block diag}[N_{\ell 1}, \cdots, N_{\ell \mu}] \in \mathbb{R}[s]^{\nu \times n_i},$$

42 $$N_r := \text{block diag}[N_{r1}, \cdots, N_{r\mu}] \in \mathbb{R}[s]^{n_o \times \nu},$$

it follows that, by (8)-(13), (26)-(29), and (38)-(42), the open-loop transfer function G of Σ, given by (13), has a r.ℓ.f. (N_r, D, N_ℓ), i.e.,

43 $$G = N_r D^{-1} N_\ell,$$

which is generated by the open-loop PMD $\mathcal{D}_{ye} := [D, N_\ell, N_r, 0]$ of Σ described in the time domain by

44 $$D(p)\xi(t) = N_\ell(p)e(t)$$

$$t \geq 0.$$

45 $$y(t) = N_r(p)\xi(t)$$

Moreover, we have

46 Fact. Under Assumptions (3), (19), and (25) the open-loop PMD $\mathcal{D}_{ye}$ of Σ given by (44)-(45) is well formed and has no unstable hidden modes.

Proof. From (38)-(43) and (13) and Fact 32 it follows that D^{-1}, $N_r D^{-1}$, $D^{-1} N_\ell$, and $G = N_r D^{-1} N_\ell$ are proper rational matrices. Hence, by Theorem 3.3.1.52, $\mathcal{D} := [D, N_\ell, N_r, 0]$ is well formed. Moreover, by (38)-(43) and Fact 32, $\mathrm{rk}[D \,\vdots\, N_\ell](s) = \nu$, $\forall s \in \mathbb{C}_+$ and $\mathrm{rk}\begin{bmatrix} D \\ N_r \end{bmatrix}(s) = \nu$, $\forall s \in \mathbb{C}_+$. Hence, by Exercise 3.3.2.47, $\mathcal{D} := [D, N_\ell, N_r, 0]$ has no $\mathbb{C}_+$-decoupling zeros. ∎

From equations (14)-(15) and (43)-(45) we obtain a closed-loop input-output PMD $\mathcal{D}_{yu} := [D_g, N_\ell, N_r, 0]$ of Σ described in the time domain by

50 $$D_g(p)\xi(t) = N_\ell(p)u(t)$$

$$t \geq 0,$$

51 $$y(t) = N_r(p)\xi(t)$$

where in the frequency domain

52 $$D_g := D - N_\ell F N_r \in \mathbb{R}[s]^{\nu\times\nu}.$$

Moreover, we also obtain a closed-loop input-error PMD $\mathcal{D}_{eu} := [D_g, N_\ell, FN_r, I]$ of Σ described in the time domain by

53 $$D_g(p)\xi(t) = N_\ell(p)u(t)$$

$$t \geq 0,$$

54 $$e(t) = FN_r(p)\xi(t) + u(t)$$

where D_g is given by (52).

The transfer functions of the PMDs $\mathcal{D}_{yu}$ and $\mathcal{D}_{eu}$ are the closed-loop transfer functions H_{yu} and H_{eu} of Σ given by (16)-(18); indeed, in the frequency domain we have

55 $$H_{yu} = N_r\, D_g^{-1}\, N_\ell \in \mathbb{R}(s)^{n_o \times n_i}$$

and

56 $$H_{eu} = FN_r\, D_g^{-1}\, N_\ell + I \in \mathbb{R}(s)^{n_i \times n_i}.$$

We now have

57 Fact. Under Assumptions (3), (19), and (25) the closed-loop PMDs $\mathcal{D}_{yu}$ and $\mathcal{D}_{eu}$ are well formed.

Proof. (a) We show that $D_g \in \mathbb{R}[s]^{\nu\times\nu}$, given by (52), has a proper inverse. By Fact 46 we have that G has an int. pr. r.ℓ.f. (N_r, D, N_ℓ) whence by Theorem 3.4.40

58 $$D = D_{cr}\, D_-\, D_{rr},$$

where $D_- \in \mathbb{R}(s)^{\nu\times\nu}$ is biproper and D_{cr} and D_{rr} are polynomial matrices with D_{cr} column-reduced and D_{rr} row-reduced s.t.

59 $$N_r\, D_{rr}^{-1} \text{ and } D_{cr}^{-1}\, N_\ell \text{ are proper rational matrices.}$$

Hence by (52) and (58)-(59),

60 $$\begin{aligned} D_g^{-1} &= (D_{cr}D_-D_{rr} - N_\ell F N_r)^{-1} \\ &= D_{rr}^{-1}(D_- - D_{cr}^{-1}N_\ell F N_r D_{rr}^{-1})^{-1} D_{cr}^{-1}, \end{aligned}$$

where D_{rr}^{-1} and D_{cr}^{-1} are proper. Moreover,

61 $$D_- - (D_{cr}^{-1}\, N_\ell)\, F(N_r\, D_{rr}^{-1}) \text{ is biproper.}$$

Indeed, by (58)-(59) it is already proper, and, using in addition (43) and (20),

$$\det(D_- - D_{cr}^{-1}N_\ell F N_r D_{rr}^{-1})(\infty) = \det D_-(\infty)\det(I - FG)(\infty) \neq 0.$$

Hence (61) follows. Now all RHS factors of (60) are proper whence D_g^{-1} is proper.

(b) The matrices $N_r\, D_g^{-1}$ and $D_g^{-1}\, N_\ell$ are proper: indeed, in

$$N_r D_g^{-1} = (N_r\, D_{rr}^{-1})(D_- - D_{cr}^{-1}N_\ell F N_r D_{rr}^{-1})^{-1} D_{cr}^{-1}$$

and

$$D_g^{-1}N_\ell = D_{rr}^{-1}(D_- - D_{cr}^{-1}N_\ell F N_r D_{rr}^{-1})^{-1}(D_{cr}^{-1}N_\ell),$$

all RHS factors are proper by (58), (59), and (61).

(c) By Assumption WP it follows that the transfer functions H_{yu} and H_{eu} are proper.

Hence in view of (a)-(c) and the fact that F is constant, it follows that the PMDs $\mathcal{D}_{yu}$ and $\mathcal{D}_{eu}$ are well formed, (Theorem 3.3.1.52). ∎

64 <u>Characteristic Polynomial and Exponential Stability of Interconnected System</u> Σ. Under Assumptions IS, WP, and PMD, by equations (14)-(15) and (43)-(45), and in view of Fig. 3 and Fact 57, the interconnected system Σ is a <u>well-formed</u> PMD $\mathcal{D}_g$ which maps the input u into (y, e, Fy) and is described in the time domain by

(a) the pseudo-state equation

50 $$D_g(p)\xi(t) = N_\ell(p)u(t) \qquad t \geq 0,$$

where D_g is given by (52);

(b) three readout equations given by

51 $$y(t) = N_r(p)\xi(t)$$

54 $$e(t) = FN_r(p)\xi(t) + u(t) \qquad t \geq 0.$$

65 $$Fy(t) = FN_r(p)\xi(t)$$

Notice that det D_g with $D_g \in \mathbb{R}[s]^{\nu\times\nu}$ is the characteristic polynomial of $\mathcal{D}_g$ according to Definition 3.2.1.15. Hence the following natural definitions.

66 Definitions. Consider the interconnected system Σ under Assumptions IS, WP, PMD, and its PMD $\mathcal{D}_g$ described by equations (50)-(52) and (54), (65).
(a)

67 We call the pseudo-state of Σ the pseudo-state $\xi(\cdot)$ of $\mathcal{D}_g$.

(b)

68 We call the characteristic polynomial (char. poly.) of Σ the characteristic polynomial of $\mathcal{D}_g$, namely

69 $$\chi(s) = \det D_g(s),$$

where $D_g \in \mathbb{R}[s]^{\nu\times\nu}$ with $D_g := D - N_\ell F N_r$.

(c)

70 We say that $\lambda \in \mathbb{C}$ is an eigenvalue of Σ iff λ is an eigenvalue of $\mathcal{D}_g$ or equiv., λ is a root of χ, (these λ's are often called "closed-loop eigenvalues" of Σ); we say that $\lambda \in \mathbb{C}$ is a decoupling zero of Σ iff λ is a decoupling zero of $\mathcal{D}_g$.

(d)

71 We say that Σ is exponentially stable (exp. stable) if and only if $\mathcal{D}_g$ is exp. stable. ∎

72 Comment. Observe that every definition is obtained by replacing $\mathcal{D}_g$ by Σ.

Since the PMD $\mathcal{D}_g$ is well formed, we obtain immediately by Theorem 3.3.2.18:

73 Theorem [Exp. stable interconnected system Σ]. Consider interconnected system Σ under Assumptions IS, WP, PMD and its characteristic polynomial $\chi(\cdot)$ given by (69).
U.t.c.

Σ is exp. stable

if and only if

$$Z[\chi] \subset \mathring{\mathbb{C}}_-$$

or, equiv.

Σ has no $\mathbb{C}_+$-eigenvalues. ∎

74 Comment. Notice that, because Σ is a PMD $\mathcal{D}_g$ mapping u into (y, e, Fy) with pseudo-state ξ (see equations (50)-(52), (54) and (65)), exponential stability of Σ implies that $\xi(\cdot)$ and $(y, e, Fy)(\cdot)$ have properties as described in Theorem 3.3.2.19: in particular, exponentially decreasing z-i trajectories and asymptotical inheritance by $(y, e, Fy)(\cdot)$ of the properties of $u(\cdot)$ as $t \to \infty \cdots$.

75 Remark. From the analysis above it is clear that interconnected system Σ is a well-formed PMD $\mathcal{D}_g$, described by (50) - (52), (54), and (65) under Assumptions IS, WP, and

76 Assumption WF. For k = 1, 2, $\cdots$, μ each subsystem transfer function $G_k \in \mathbb{R}_p(s)^{n_{ok} \times n_{ik}}$ has a right-left fraction $(N_{rk}, D_k, N_{\ell k})$ generated by a well-formed PMD $\mathcal{D}_k = [D_k, N_{\ell k}, N_{rk}, 0]$ according to (26)-(29).

As a consequence Definitions 66 and Theorem 73 are valid under the Assumptions IS, WP, and WF.

This is important in the feedback compensator design problem (see below). ∎

We shall now relate the exp. stability of Σ to that of the input-output transfer function H_{yu} of Σ, given by (16)-(17). We first note that H_{yu} is the transfer function of the input-output PMD $\mathcal{D}_{yu} = [D_g, N_\ell, N_r, 0]$ of Σ and have:

77 Fact. Consider an interconnected system Σ under Assumptions IS, WP, and PMD. Let $\mathcal{D}_{yu} = [D_g, N_\ell, N_r, 0]$ be the input-output PMD of Σ given by (50)-(52).

U.t.c.

(a) $\mathcal{D}_{yu}$ is well formed,

(b) $\mathcal{D}_{yu}$ has no unstable hidden modes.

Proof. In view of Fact 57 we must only show that (b) holds. Observe now that the open-loop PMD $\mathcal{D}_{ye} = [D, N_\ell, N_r, 0]$ of Σ has no $\mathbb{C}_+$-decoupling zeros according to Fact 46, or equiv. by Exercise 3.3.2.47

78 $$\text{rk}[D \mid N_\ell](s) = \nu \quad \forall s \in \mathbb{C}_+ \quad \text{and} \quad \text{rk}\begin{bmatrix} D \\ \hline N_r \end{bmatrix}(s) = \nu \quad \forall s \in \mathbb{C}_+.$$

Now by (52),

79 $$[D_g \mid N_\ell] = [D \mid N_\ell]\left[\begin{array}{c|c} I & 0 \\ \hline -FN_r & I \end{array}\right] \quad \text{and} \quad \begin{bmatrix} D_g \\ \hline N_r \end{bmatrix} = \left[\begin{array}{c|c} I & -N_\ell F \\ \hline 0 & I \end{array}\right]\begin{bmatrix} D \\ \hline N_r \end{bmatrix},$$

where the transforming matrices in the RHS are unimodular polynomial matrices. Hence by (78)-(79)

80 $$\text{rk}[D_g \mid N_\ell](s) = \nu \quad \forall s \in \mathbb{C}_+ \quad \text{and} \quad \text{rk}\begin{bmatrix} D_g \\ \hline N_r \end{bmatrix}(s) = \nu \ \forall s \in \mathbb{C}_+.$$

Therefore, by Exercise 3.3.2.47, $\mathcal{D}_{yu}$ has no $\mathbb{C}_+$-decoupling zeros. ∎

81 Exercise. Consider interconnected system Σ under Assumptions IS, WP, and PMD. Let $\mathcal{D}_{eu} = [D_g, N_\ell, FN_r, I]$ be the input-error PMD of Σ, given by (53)-(54), with transfer function H_{eu} (see equations (17) and (56)).

(a) $\mathcal{D}_{eu}$ is well formed.

(b) $\mathcal{D}_{eu}$ has no unstable hidden modes iff $\mathcal{D}_{eu}$ has no $\mathbb{C}_+$-output-decoupling zeros, or equiv.

82 $$\text{rk}\begin{bmatrix} D \\ \hline FN_r \end{bmatrix}(s) = \nu \quad \forall s \in \mathbb{C}_+.$$

83 Comments. (a) In view of (78), criterion (82) is satisfied if the feedback gain matrix F of Σ (see (12)) is left invertible. (Roughly, the feedback must be "complete": Fy must determine y completely.) Note that (78) does not imply (82): e.g., take F = 0 and the open-loop PMD $\mathcal{D}_{ye} = [D, N_r, N_\ell, 0]$ not exp. stable.

(b) Criterion (82) is also satisfied if the open-loop PMD $\mathcal{D}_{ye}$ is exp. stable.

In view of Theorem 73, Fact 77, Exercise 81, and Definition 3.3.2.31 of exp. stable transfer functions, we now have by Theorem 3.3.2.40:

84 Theorem [Exp. stability of Σ by transfer functions]. Consider interconnected system Σ under Assumptions IS, WP, and PMD. Let H_{yu}, resp. $H_{eu} \in \mathcal{E}(\mathbb{R}_p(s))$, be the input-output and input-error transfer functions of Σ given by (16)-(18). Consider also the input-error PMD $\mathcal{D}_{eu} = [D_g, N_\ell, FN_r, I]$

of Σ given by (53)-(54).

U.t.c.

(a)

85 $\quad \Sigma$ is exp. stable

$\Leftrightarrow$

86 $\quad H_{yu}$ is exp. stable

or equiv.

87 $\quad \mathcal{P}[H_{yu}] \subset \mathring{\mathbb{C}}_-$.

(b) Iff, in addition, $\mathcal{D}_{eu}$ has no $\mathbb{C}_+$-output-decoupling zeros, or equiv.

82 $$\operatorname{rk}\begin{bmatrix} D \\ \hline FN_r \end{bmatrix}(s) = \nu \quad \forall s \in \mathbb{C}_+,$$

then

85 $\quad \Sigma$ is exp. stable

$\Leftrightarrow$

88 $\quad H_{eu}$ is exp. stable

or equiv.

89 $\quad \mathcal{P}[H_{eu}] \subset \mathring{\mathbb{C}}_-$. ∎

90 Comments. (a) Using the classification of eigenvalues of the PMD $\mathcal{D}_{yu} = [D_g, N_\ell, N_r, 0]$ of Σ, especially equation (3.2.4.18), one has that, under the conditions of Theorem 84, the characteristic polynomial of Σ (see (69)) reads

91 $$\chi(s) = \det D_g(s) = \delta(s)\pi(s),$$

where $\delta(\cdot)$ and $\pi(\cdot)$ are polynomials such that

(i) $\pi(p) = 0$ iff $p \in \mathcal{P}[H_{yu}]$,
(ii) $\delta(z) = 0$ iff $z \in \mathbb{C}$ is a decoupling zero of $\mathcal{D}_{yu}$,
(iii) $Z[\delta] \subset \mathring{\mathbb{C}}_-$.

As a consequence $Z[\chi] \subset \mathring{\mathbb{C}}_-$ iff $Z[\pi] \subset \mathring{\mathbb{C}}_-$, which explains (85)⇔(86)⇔(87).
(b) The equivalence (85)⇔(88)⇔(89) is satisfied if the gain matrix F has a left inverse in $\mathbb{R}^{n_o \times n_i}$ ("complete feedback") or if the open-loop PMD $\mathcal{D}_{ye}$ is exp. stable.

92 Corollary. Consider an interconnected system Σ under Assumptions IS and WP, with Assumption PMD strengthened to

93 Assumption PMD': The interconnected system Σ is such that $\forall k = 1, 2, \cdots, \mu$ each subsystem transfer function $G_k \in \mathbb{R}_p(s)^{n_{ok} \times n_{ik}}$ has a right-left-coprime fraction $(N_{rk}, D_k, N_{\ell k})$ generated by a well-formed and minimal PMD $\mathcal{D}_k := [D_k, N_{\ell k}, N_{rk}, 0]$ described by equations (26)-(29).

U.t.c.
(a) The open-loop PMD $\mathcal{D}_{ye} := [D, N_\ell, N_r, 0]$ and the input-output PMD $\mathcal{D}_{yu} = [D_g, N_\ell, N_r, 0]$ of Σ, given by (43)-(45) and (50)-(52), are well formed and minimal, or equiv. (N_r, D, N_ℓ) and (N_r, D_g, N_ℓ) are int. pr. r.ℓ.c.f.'s of G and H_{yu} given by (43) resp. (17).
(b) Equation (91) reads

94 $\chi(\cdot) \sim \pi(\cdot)$

or equiv. the characteristic polynomial $\chi(\cdot)$ of Σ and $\pi(\cdot)$ are equal modulo a nonzero constant s.t.

95 $p \in P[H_{yu}] \Leftrightarrow \chi(p) = 0,$

i.e., p is a pole of H_{yu} iff p is an eigenvalue of Σ. ∎

96 Exercise. Prove Corollary 92.

4.3. Feedback System Exponential Stability

Consider the feedback system Σ of Fig. 1.

1 Assumptions. In feedback system Σ of Fig. 1, P is the plant and C is the compensator s.t.

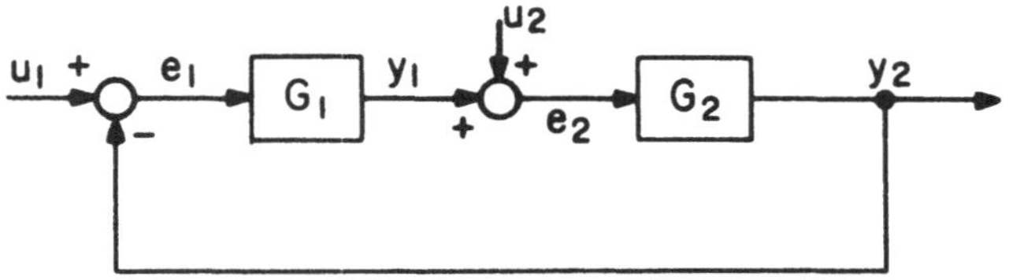

Fig. 1. The feedback system Σ under consideration.

2 $P \in \mathbb{R}_p(s)^{n_o \times n_i}$ has an int. pr. ℓ.c.f. $(D_{p\ell}, N_{p\ell})$ and an int. pr. r.c.f. (N_{pr}, D_{pr});

3 $C \in \mathbb{R}_p(s)^{n_i \times n_o}$ has an int. pr. ℓ.c.f. $(D_{c\ell}, N_{c\ell})$ and an int. pr. r.c.f. (N_{cr}, D_{cr});

4 $\det(I_{n_o} + PC)(\infty) = \det(I_{n_i} + CP)(\infty) \neq 0$.

5 <u>Comments</u>. (a) Assumption (4) is satisfied if P is strictly proper.

(b) In most applications Assumptions (2)-(3) are satisfied with the left denominators row-reduced and the right denominators column-reduced.

6 <u>Theorem</u> [Feedback system stability]. Consider feedback system Σ given by Fig. 1 under the Assumptions 1.

U.t.c.

(a) Σ is an interconnected system satisfying Assumptions IS, WP, and PMD of Sec. 4.2.

(b) The input-output transfer function H_{yu} of Σ reads

$$H_{yu} = \begin{bmatrix} H_{y_1u_1} & H_{y_1u_2} \\ H_{y_2u_1} & H_{y_2u_2} \end{bmatrix}$$

7

$$= \begin{bmatrix} C(I + PC)^{-1} & -CP(I + CP)^{-1} \\ PC(I + PC)^{-1} & P(I + CP)^{-1} \end{bmatrix}.$$

(c) The <u>characteristic polynomial</u> of Σ, defined in (4.2.69), has four <u>equivalent expressions</u> (equal modulo a nonzero constant):

8 $$\chi(s) \sim \det\,[D_{c\ell}\, D_{pr} + N_{c\ell}\, N_{pr}](s),$$

9 $$\sim \det\,[D_{p\ell}\, D_{cr} + N_{p\ell}\, N_{cr}](s),$$

10 $$\sim \det\left[\begin{array}{c|c} D_{cr} & N_{pr} \\ \hline -N_{cr} & D_{pr} \end{array}\right](s),$$

11 $$\sim \det\left[\begin{array}{c|c} D_{c\ell} & N_{c\ell} \\ \hline -N_{p\ell} & D_{p\ell} \end{array}\right](s).$$

(d)

12 $$p \in P[H_{yu}] \Leftrightarrow \chi(p) = 0.$$

(e)

13 $$\Sigma \text{ is exp. stable} \Leftrightarrow Z[\chi] \subset \mathring{\mathbb{C}}\ .$$

(f) If $H_{eu} : u = (u_1, u_2) \mapsto e = (e_1, e_2)$ is the input-error transfer function of Σ, then

14 $$p \in P[H_{yu}] \Leftrightarrow p \in P[H_{eu}].$$ ■

15 <u>Comment</u>. The four expressions of the char. poly. of Σ reflect the four possible fractional representations of P and C in (2)-(3).

16 <u>Proof of Theorem 6</u>.

(a) Identify $G_1 := C$ and $G_2 := P$ as subsystem transfer matrices, with $n_{o1} = n_i$, $n_{i1} = n_o$, $n_{o2} = n_o$, and $n_{i2} = n_i$. From Fig. 1 it follows that Σ satisfies Assumption IS with F and G in (4.2.12)-(4.2.13) reading

17 $$F = \left[\begin{array}{c|c} 0 & -I \\ \hline I & 0 \end{array}\right], \quad G = \left[\begin{array}{c|c} C & 0 \\ \hline 0 & P \end{array}\right].$$

Assumption WP is satisfied because of (4); indeed, by (17),

$$\det[I - FG](\infty) = \det\left[\begin{array}{c|c} I & P \\ \hline -C & I \end{array}\right](\infty) = \det(I + PC)(\infty) = \det(I + CP)(\infty) \neq 0.$$

Finally, Assumption PMD is satisfied by assumptions (2)-(3). Hence Σ is an interconnected system satisfying Assumptions IS, WP, and PMD.

(b) Exercise: follows by straightforward computation.

(c) We shall compute (8) in detail. Choose for $G_1 := C$ an int. pr. ℓ.c.f. $(D_{c\ell}, N_{c\ell})$ and for $G_2 := P$ an int. pr. r.c.f. (N_{pr}, D_{pr}). Then, in (4.2.40)-(4.2.42), we have

18 $$D = \text{block diag}[D_{c\ell}, D_{pr}],$$

19 $$N_\ell = \text{block diag}[N_{c\ell}, I],$$

20 $$N_r = \text{block diag}[I, N_{pr}].$$

As a consequence, (4.2.52) reads

21 $$D_g = D - N_\ell F N_r = \left[\begin{array}{c|c} D_{c\ell} & N_{c\ell}N_{pr} \\ \hline -I & D_{pr} \end{array}\right]$$

Hence in (4.2.69) the characteristic polynomial of Σ reads

$$\chi(s) = \det D_g(s) = \det\left[\begin{array}{c|c} D_{c\ell} & D_{c\ell}D_{pr} + N_{c\ell}N_{pr} \\ \hline -I & 0 \end{array}\right]$$

$$\sim \det[D_{c\ell}\, D_{pr} + N_{c\ell}\, N_{pr}](s).$$

Hence (8) holds.

Expressions (9)-(11) are computed similarly by choosing for $G_1 := C$, $G_2 := P$ resp. (a) an int. pr. r.c.f. and an int. pr. ℓ.c.f., (b) an int. pr. r.c.f. and an int. pr. r.c.f., (c) an int. pr. ℓ.c.f. and an int. pr. ℓ.c.f.. We notice now that, in the computations above, we worked under assumptions (2)-(4), whence Σ is an interconnected system satisfying Assumptions IS, WP, and PMD'(see Corollary 4.2.92). Hence

any expression on the RHS of (8)-(11) is a denominator
22 determinant of r.ℓ.c.f. (N_r, D_g, N_ℓ) of the input-output transfer function H_{yu} of Σ.

Expressions (8)-(11) are therefore equivalent for the reasons sketched below:

1) Any r.ℓ.c.f. (N_r, D_g, N_ℓ) of H_{yu} can be converted into a r.c.f. $(N_r\bar{N}_r, \bar{D}_r)$ of H_{yu}, where D_g and $\bar{D}_r$ have the same nonunity invariant polynomials. (Hint: Use generalized Bezout identities for the coprime fractions $N_rD_g^{-1}$ and $D_g^{-1}N_\ell$ and repeat the arguments in the proof of Theorem 2.4.4.12.)

2) According to Theorem 2.4.2.41, all denominators of any coprime fraction of H_{yu} have the same nonunity invariant polynomials; in particular their determinants are equivalent.

(d) This is a consequence of (22) and Theorem 3.4.3.

(e) This is a consequence of (c) and Theorem 4.2.73.

(f) Notice that the feedback matrix $F \in \mathbb{R}^{(n_o+n_i)\times(n_o+n_i)}$, given by (17), is invertible. Therefore, (14) is a consequence of (4.2.18). ∎

23 <u>Exercise</u>. Consider Σ defined in (1)-(4). Let ζ be a zero of $\chi_p = \det D_{pr}$. Let $N_{c\ell}(\zeta)$ be full rank. Show that (a) in the single-input single-output case it is always true that $\chi(\zeta) \neq 0$ (where χ is given by (8)); (b) in the multi-input multi-output case, it <u>may</u> happen that $\chi(\zeta) = 0$. (Roughly speaking, with pole-zero cancellations between P and C ruled out, (a) for the single-input single-output case, a plant pole ζ <u>can never</u> be a closed-loop eigenvalue; (b) in the multi-input multi-output case, a plant pole ζ <u>may</u> be a closed-loop eigenvalue.) ∎

In feedback compensator design we may encounter a compensator which is not a coprime fraction.

26 <u>Corollary</u> [C not a coprime fraction]. Under the assumptions of Theorem 6, where Assumption (3) is replaced by

27 $C \in \mathbb{R}_p(s)^{n_i\times n_o}$ has an int. pr. ℓ.f. $(D_{c\ell}, N_{c\ell})$ or an int. pr. r.f. (N_{cr}, D_{cr}),

we have that:

(a) Feedback system Σ, shown in Fig. 1, is an interconnected system Σ satisfying assumptions IS, WP, and WF of Sec. 4.2 (see Remark 4.2.75).
(b) Σ has input-output- and input-error transfer functions H_{yu} resp. H_{eu}, given by (7) and

28 $$H_{eu} = I + FH_{yu}$$

with

17 $$F = \begin{bmatrix} 0 & -I \\ I & 0 \end{bmatrix}.$$

(c) The four possible choices of fractions for P and C in (2) and (27) define four well-formed closed-loop PMDs $\mathcal{D}_g$ of Σ as in (4.2.64) et seq.; each choice generates a char. poly. χ for Σ, given by an expression on the RHS of (8)-(11); furthermore, Σ is exp. stable iff $Z[\chi] \subset \mathring{\mathbb{C}}_-$.
(d) If P has an int. pr. r.c.f. (N_{cr}, D_{cr}) and C has an int. pr. ℓ.f. $(D_{c\ell}, N_{c\ell})$, then the following holds: Let $L \in \mathcal{E}(\mathbb{R}[s])$ be any g.c.ℓ.d. of $(D_{c\ell}, N_{c\ell})$, whence

29 $$D_{c\ell} = L\bar{D}_{c\ell} \quad \text{and} \quad N_{c\ell} = L\bar{N}_{c\ell}$$

for some $\bar{D}_{c\ell}, \bar{N}_{c\ell} \in \mathcal{E}[\mathbb{R}[s])$ s.t.

30 $$\mathrm{rk}[\bar{D}_{c\ell} \,\vdots\, \bar{N}_{c\ell}](s) = n_i \quad \forall s \in \mathbb{C}.$$

Then the char. poly. of Σ reads

31 $$\chi(s) = \det[D_{c\ell}D_{pr} + N_{c\ell}N_{pr}](s)$$
$$= \det L(s) \det[\bar{D}_{c\ell}D_{pr} + \bar{N}_{c\ell}N_{pr}](s),$$

where

32 $$p \in \mathcal{P}[H_{yu}] \Leftrightarrow p \in \mathcal{P}[H_{eu}] \Leftrightarrow \det[\bar{D}_{c\ell}D_{pr} + \bar{N}_{c\ell}N_{pr}](p) = 0$$

and

33 $$z \text{ is a decoupling zero of } \Sigma \Leftrightarrow \det L(z) = 0$$
$$\Leftrightarrow z \text{ is a decoupling zero of C.}$$

34 Exercise. Prove Corollary 26. (For (33) note that since F is invertible, z is a decoupling zero of Σ iff z is a decoupling zero of the open-loop PMD $[D, N_\ell, N_r, 0]$ given in (4.2.44)-(4.2.45).)

35 Remark. In the spirit of part (c) of Corollary 26, suppose that both P and C are specified by not necessarily coprime fractions. Then it is easy to check that the characteristic polynomial of Σ is still given by the corresponding expression in the RHS of (8)-(11).

4.4. Special Properties of Feedback Systems

Consider feedback system Σ, shown in Fig. 4.3.1, under the assumptions (4.3.1) with $P = N_{pr} D_{pr}^{-1} \in E(\mathbb{R}_{p,o}(s))$ and $C = D_{c\ell}^{-1} N_{c\ell}$.

1 (a) Note that in (4.3.8), the char. poly. of Σ satisfies

$$\begin{aligned}\chi_\Sigma(s) &= \det[D_{c\ell}D_{pr} + N_{c\ell}N_{pr}](s)\\ &= \det(I + PC)(s)\cdot\det D_{c\ell}(s)\cdot\det D_{pr}(s).\end{aligned}$$

where $\det D_{c\ell} =: \chi_c$ and $\det D_{pr} =: \chi_p$ are the compensator- and plant-char. poly.'s. Hence

2 $$\det(I + PC)(s) = \chi_\Sigma(s)\Big/\big(\chi_c(s)\ \chi_p(s)\big),$$

i.e., the determinant of the return difference matrix is the ratio of the closed-loop char. poly. over the open-loop char. poly. of Σ.

This relation is useful for establishing a graphical test for stability (Nyquist criterion) (see Exercise 11 below).

3 (b) Note that if $C = k\,D_{c\ell}^{-1} N_{c\ell}$, where $k \in \mathbb{R}$ is an adjustable loop gain, then the char. poly. of Σ satisfies

4 $$\begin{aligned}\chi(s, k) &= \det[D_{c\ell}D_{pr} + kN_{c\ell}N_{pr}](s)\\ &= k^{n_i}\det[\tfrac{1}{k} D_{c\ell}D_{pr} + N_{c\ell}N_{pr}](s).\end{aligned}$$

Suppose now that k is very large; then the polynomial in (4) has coefficients which are close to the coefficients of the polynomial $\det[N_{c\ell}N_{pr}]$.
Therefore, by the continuous dependence of the zeros of a polynomial on its

coefficients, the char. poly. $\chi(s, k)$ has, for k large, at least one zero (say z_i) close to every zero of $\det[N_{c\ell}N_{pr}]$ (say z_i'). Furthermore, for all such zeros of $\chi(s, k)$, $|z_i - z_i'| \to 0$ as $k \to \infty$. Therefore, if the plant P has $\mathring{\mathbb{C}}_+$-zeros, then as the feedback becomes tighter (i.e., $k \to \infty$), a closed-loop eigenvalue tends to each of the $\mathring{\mathbb{C}}_+$-zeros of the plant, whence for k sufficiently large Σ will be unstable.

5 Conclusion. A plant with $\mathring{\mathbb{C}}_+$-zeros imposes a bound on the amount of loop gain that a feedback system can tolerate.

6 Exercise. Show that the statement (5) may be false if $\mathring{\mathbb{C}}_+$ is replaced by $\mathbb{C}_+$. (Hint: Let k_1, k_2, p_1, p_2 be positive numbers; let $p(s) = k_1 s/(s + p_1)^2$, $c(s) = k_2/(s + p_2)$. With k_1 finite but $k_2 \to \infty$, show that the zeros of $\chi(s, k_2)$ remain in $\mathring{\mathbb{C}}_-$. Give in addition a proof based on the Nyquist criterion.)

8 (c) In applications the usual I/O map of Σ is given by

9
$$H_{y_2u_1} = PC(I + PC)^{-1} = P(I + CP)^{-1}C$$
$$= N_{pr}(D_{c\ell}D_{pr} + N_{c\ell}N_{pr})^{-1}N_{c\ell}.$$

Now, if Σ is exp. stable, then, by (4.3.8), $Z[\det(D_{c\ell}D_{pr} + N_{c\ell}N_{pr})] \subset \mathring{\mathbb{C}}_-$ and there cannot be any cancellation of a right nonunimodular common factor with $\mathbb{C}_+$-eigenvalues between N_{pr} and the denominator of $H_{y_2u_1}$ in (9). Hence we have

10
$$Z[P] \cap \mathbb{C}_+ \subset Z[H_{y_2u_1}],$$

i.e., feedback cannot remove the $\mathbb{C}_+$-zeros of the plant.

The following exercise completes the information given at the end of (a).

11 Exercise [Multivariable Nyquist test]. Consider the system Σ specified by the assumptions (4.3.1).

(a) Assume that neither χ_p nor χ_c have $j\omega$-axis zeros. Let n_p (n_c) denote the number of $\mathbb{C}_+$-zeros of χ_p (χ_c, resp.). Let R be real, positive, and arbitrarily large. Define the oriented curve D_R be the straight-line segment from $(0, -jR)$ to $(0, jR)$ and the right half-plane semicircle joining these points. Note that D_R is traversed clockwise. Let $\phi : \mathbb{C}_+ \to \mathbb{C}$ be analytic on D_R; then the Nyquist diagram of ϕ is, by definition, the oriented curve $\phi(D_R)$, namely the map of D_R by ϕ. Use the principle of the argument to show that

$Z[\chi_\Sigma(s)] \subset \mathring{\mathbb{C}}_- \Leftrightarrow$ the Nyquist diagram of $\det[I + PC](s)$ does not go through the origin and encircles it $n_p + n_c$ times counterclockwise.

(c) In case χ_p or χ_c have $j\omega$-axis zeros, show that for the statement above to remain true, the contour D_R must be indented on the left (see Exercise 4.3.23).

The Nyquist criterion is the basis for the following robust stability result.

12 Remark [Robust stability] [Doy.1]. Call $\Sigma(P, C)$ the system described by (4.3.1)-(4.3.3) and $P \in \mathbb{R}_{p,o}(s)^{n_o \times n_i}$. Model plant perturbations as follows: P is replaced by $(I + M)P$, resulting in a perturbed system $\Sigma((I + M)P, C)$. As in (4.3.2), $(I + M)P$ is specified by an int. proper coprime fraction, hence $\chi_{(I+M)P}$ is well defined. The size of the perturbation M is bounded by a given continuous tolerance curve $\ell(\cdot) : \mathbb{R}_+ \to \mathbb{R}_+$, s.t. $\ell(\omega) > 0$, $\forall \omega \in \mathbb{R}_+$, and $\exists k \in \mathbb{N}$ s.t. $\ell(\cdot)\omega^k > 1$ for ω sufficiently large.

We say that $M \in \mathcal{M}$ iff $M \in \mathbb{R}(s)^{n_o \times n_o}$ is s.t.

(a) $(I + M)P \in \mathbb{R}_{p,o}(s)^{n_o \times n_i}$;

(b) $\sigma_{max}[M(j\omega)] < \ell(\omega)$ $\forall \omega \in \mathbb{R}_+$, where $\sigma_{max}[\cdots]$ denotes the maximal singular value;

(c) the number of $\mathbb{C}_+$-zeros of $\chi_{(I+M)P}$ equals the number of $\mathbb{C}_+$-zeros of χ_P, counting multiplicities. (Equivalently, P and $(I + M)P$ have the same number of $\mathbb{C}_+$-poles, with multiple poles counted according to their Mcmillan degree [Kai.1, Sec. 6.5].)

The following can then be proven:

Let $\Sigma(P, C)$ be exp. stable

U.t.c.

$$\forall M \in \mathcal{M}, \ \Sigma\big((I + M)P, C\big) \text{ is exp. stable}$$

$\Leftrightarrow$

$$\sigma_{max}[PC(I + PC)^{-1}(j\omega)] \le 1/\ell(\omega), \ \forall \omega \in \mathbb{R}_+.$$

13 Comment. A remarkable feature of the result of Remark 12 is that the three transfer functions P, C, and (I + M)P may be unstable. In fact, by suitable choice of M, P, and (I+M)P may have no zeros and no poles in common!

14 Exercise. Let

$$P = \begin{bmatrix} \dfrac{1}{s-1} & \dfrac{s}{s^2+s+1} \\ 0 & \dfrac{s-2}{(s+1)(s+2)} \end{bmatrix}$$

Find an $M \in \mathbb{R}_p(s)^{2\times 2}$ such that $\mathbb{C}_+$-zero of P and the $\mathbb{C}_+$-pole of P are moved to $2 + \Delta_2$ and to $1 + \Delta_1$ respectively. ∎

Chapter 5. Single-Input Single-Output Systems

5.1. Introduction

The purpose of this chapter is to treat, in the simple context of single-input single-output feedback systems [Chen.1], a number of design problems that will be treated in the multivariable case in Chapters 6, 7, and 8.

In Sect. 2 we study, for a given plant, the problem of designing a two-input one-output compensator. Such compensator has the advantage that it allows independent adjustment of the closed-loop dynamics and of the I/O map. The realization of the compensator is carefully described because it is of crucial importance in the determination of closed-loop characteristic polynomial (5.2.26). This characteristic polynomial is used in the definition of $\mathcal{U}$-stability where the closed-loop eigenvalues in the undesirable set $\mathcal{U}$ of the complex plane are ruled out (Theorem 5.2.37). The properness of the compensator is guaranteed by simple conditions (Theorem 5.2.47). Corollary 5.2.56 describes the class of all achievable I/O maps using such a compensator and the given plant. Finally, the concept of robust asymptotic tracking is carefully defined and Theorem 5.2.71 specifies the necessary and sufficient conditions that must be satisfied so that robust asymptotic tracking is achieved.

Section 3 on design contains two algorithms. The first algorithm obtains a compensator that achieves a prescribed I/O map for the closed-loop system; this algorithm is followed by comments which justify it and which describe ideas which will be useful in the multivariable case in Sec. 6.2. The second algorithm obtains a compensator which achieves robust asymptotic tracking.

5.2. Problem Statement and Analysis

We consider a linear time-invariant feedback system Σ_2 with single input, v_1, and single output, y_2, as shown in Fig. 1, where u_1, u_2, and d_o are disturbance scalar inputs (with d_o a possible disturbance at the output of the plant p), and y_1 is the scalar output of the compensator. The problem is the following. Given a plant transfer function $p \in \mathbb{R}_{p,o}(s)$, we wish to design a <u>proper</u> compensator with <u>two inputs</u>, namely v_1 and e_1, and <u>one output</u> y_1, s.t.

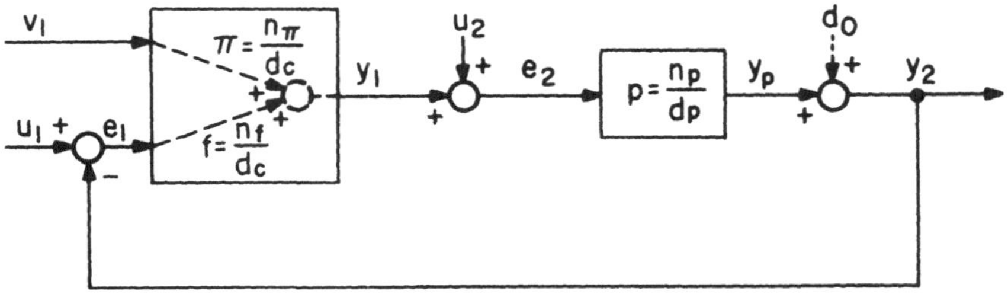

Fig. 1. The feedback system Σ_2 under consideration.

(i) system Σ_2 is stable in some sense;

(ii) the system I/O map $h_{y_2v_1}$ has, for example, a "good" step response; and

(iii) other design specifications (such as tracking, disturbance rejection, desensitization, etc.) are satisfied.

The compensator under consideration can be viewed as consisting of a precompensator with transfer function π: $v_1 \mapsto y_1$ and a feedback compensator with transfer function $f : e_1 \mapsto y_1$. In terms of realization, we let $\pi := n_\pi/d_c$ and $f := n_f/d_c$, where d_c is a <u>common denominator</u> of the rational transfer functions π and f, and n_π, $n_f \in \mathbb{R}[s]$ are the corresponding numerator, respectively. We propose to realize the two-input one-output controller using the observer canonical form [Kai.1, Fig. 2.1.9, p. 43]. More precisely, $1/d_c$ is first realized by using constant-gain feedback loops around cascaded integrators; the inputs v_1 and e_1 are then fed through appropriate constant gains to the integrator inputs to obtain n_π and n_f, respectively.

This section establishes three theorems: (a) for the <u>*U*-stability</u> of feedback system Σ_2, (b) for the existence of a <u>proper</u> compensator, and (c) for <u>robust asymptotic tracking</u> over a specified class of inputs, respectively.

1 <u>Assumptions</u>. For system Σ_2, shown in Fig. 1, we assume that:

(a) the plant is given by its transfer function

2 $p = n_p/d_p \in \mathbb{R}_{p,o}(s)$,

where n_p and d_p are coprime polynomials with $\partial[d_p] \geq 1$; moreover, p is generated by the well-formed and minimal plant PMD $\mathcal{D}_p = [d_p, 1, n_p, 0]$ with plant pseudo-state $\xi_p(\cdot)$ and given by

3 $$d_p(p)\xi_p(t) = e_2(t)$$
$$t \geq 0.$$
4 $$y_2(t) = n_p(p)\xi_p(t)$$

(b) the compensator is given by the transfer function

5 $$c = [\pi \mid f] = [n_\pi \mid n_f]/d_c \in \mathbb{R}_p(s)^{1\times 2},$$

where n_π, n_f and $d_c \in \mathbb{R}[s]$; moreover, c is generated by the well-formed compensator PMD $\mathcal{D}_c = [d_c, [n_\pi \mid n_f], 1, 0]$, with compensator pseudo-state $\xi_c(\cdot)$ and given by

6 $$d_c(p)\xi_c(t) = n_\pi(p)v_1(t) + n_f(p)e_1(t)$$
$$t \geq 0.$$
7 $$y_1(t) = \xi_c(t)$$

10 <u>Analysis</u>. Under the Assumptions 1, system Σ_2, shown in Fig. 1, is described in the time domain by a PMD $\mathcal{D}_g$, with global input $u = (u_1, u_2, v_1)^T$, pseudo-state $\xi = (\xi_p^T, \xi_c)^T$, and output $(y^T, e^T)^T$, where $y := (y_1, y_2)^T$ and $e := (e_1, e_2)^T$, by

(a) a pseudo-state equation, viz.,

11 $$D(p)\xi(t) = N_\ell(p)u(t) \qquad t \geq 0,$$

where

12 $$D = \left[\begin{array}{c|c} d_p & -1 \\ \hline n_f n_p & d_c \end{array}\right] \in \mathbb{R}[s]^{2\times 2},$$

13 $$N_\ell = \left[\begin{array}{c|c|c} 0 & 1 & 0 \\ \hline n_f & 0 & n_\pi \end{array}\right] \in \mathbb{R}[s]^{2\times 3},$$

(b) Two readout maps, viz.,

14 $$y(t) = N_r(p)\xi(t)$$

$$t \geq 0,$$

15 $$e(t) = FN_r(p)\xi(t) + Gu(t)$$

where

16 $$e(t) = Fy(t) + Gu(t) \qquad t \geq 0,$$

with

17 $$N_r = \left[\begin{array}{c|c} 0 & 1 \\ \hline n_p & 0 \end{array}\right] \in \mathbb{R}[s]^{2\times 2},$$

18 $$F = \left[\begin{array}{c|c} 0 & -1 \\ \hline 1 & 0 \end{array}\right] \in \mathbb{R}^{2\times 2}, \quad G = \left[\begin{array}{c|c|c} 1 & 0 & 0 \\ \hline 0 & 1 & 0 \end{array}\right] \in \mathbb{R}^{2\times 3}.$$

Moreover, d_o, disturbance at the output, is dynamically accounted for by replacing y_2 and u_1 resp. by $y_2 - d_o$ and $u_1 - d_o$ in the equations (11)-(18) describing the PMD $\mathcal{D}_g$ of Σ_2.

Note also that the PMD $\mathcal{D}_g$ of Σ_2 contains the global input-output PMD

19 $$\mathcal{D}_{yu} = [D, N_\ell, N_r, 0]$$

and the global input-error PMD

20 $$\mathcal{D}_{eu} = [D, N_\ell, FN_r, G],$$

with transfer functions

21 $$H_{yu} = N_r D^{-1} N_\ell \in \mathbb{R}_p(s)^{2\times 3}$$

22 $$H_{eu} = FN_r D^{-1} N_\ell + G \in \mathbb{R}_p(s)^{2\times 3},$$

where, because $F^{-1} = -F$ (see (18)),

23 $$H_{eu} = FH_{yu} + G \text{ and } H_{yu} = -FH_{eu} + FG.$$ ∎

The following fact is not unexpected in view of Fact 4.2.57.

25 Fact. Under the Assumptions 1, feedback system Σ_2 has a well-formed PMD $\mathcal{D}_g$ (11)-(18) and the same holds for the PMDs $\mathcal{D}_{yu}$ and $\mathcal{D}_{eu}$ described in (19)-(20).

Proof. In view of Theorem 3.3.1.52 and the descriptions of $\mathcal{D}_g$, $\mathcal{D}_{yu}$, and $\mathcal{D}_{eu}$ it is sufficient to show that under the Assumptions 1,

$$D^{-1},\ N_r D^{-1},\ D^{-1} N_\ell, \text{ and } H_{yu} = N_r D^{-1} N_\ell \in E(\mathbb{R}_p(s)),$$

where D, N_r, and N_ℓ are given by resp. (12), (13), and (17). Now this follows easily by inspection: for example,

$$D^{-1} = (d_c\, d_p + n_f\, n_p)^{-1} \left[\begin{array}{c|c} d_c & 1 \\ \hline -n_f n_p & d_p \end{array} \right] = (1 + fp)^{-1} \left[\begin{array}{c|c} d_p^{-1} & d_c^{-1} d_p^{-1} \\ \hline -fp & d_c^{-1} \end{array} \right],$$

where all elements in the RHS of the second equality are proper because of assumptions (2) and (5). ∎

Note also from the analysis done in (10) that the PMDs $\mathcal{D}_g$, $\mathcal{D}_{yu}$, and $\mathcal{D}_{eu}$ of Σ_2 have a common pseudo-state ξ and a common characteristic polynomial

26 $$\chi(s) = \det D(s) = (d_c\, d_p + n_f\, n_p)(s),$$

which in view of Fact 25 and Theorem 3.3.2.18 must have roots in $\mathring{\mathbb{C}}_-$ for $\mathcal{D}_g$, $\mathcal{D}_{yu}$, and $\mathcal{D}_{eu}$ to be exp. stable. Hence the following definitions make sense.

27 Definitions. Consider feedback system Σ_2, shown in Fig. 1, under the Assumptions 1. Consider also the PMD $\mathcal{D}_g$ of Σ described by (11)-(18).

(a)

28 We call pseudo-state of Σ_2 the pseudo-state $\xi = (\xi_p, \xi_c)^T$ of $\mathcal{D}_g$.

(b)

29 We call char. poly. of Σ_2 the char. poly. of $\mathcal{D}_g$, namely,

26 $$\chi(s) = \det D(s) = (d_c\, d_p + n_f\, n_p)(s).$$

(c)

30 We say that $\lambda \in \mathbb{C}$ is a (closed-loop) eigenvalue of Σ_2 iff λ is an

eigenvalue of $\mathcal{D}_g$ or equiv. λ is a root of χ.

(d)

31 We say that Σ_2 <u>is exp. stable</u> iff $\mathcal{D}_g$ is exp. stable.

From the definitions above, Fact 25 and Theorem 3.3.2.18, we now have

32 <u>Fact</u>. Consider feedback system Σ_2, shown in Fig. 1, under the Assumptions 1.

U.t.c.

Σ_2 is exp. stable

iff

the char. poly. χ of Σ_2, given by (26), satisfies

33 $$Z[\chi] \subset \mathring{\mathbb{C}}_- = \mathbb{C} \backslash \mathbb{C}_+.$$

or, equiv. iff

Σ_2 has no eigenvalues λ in $\mathbb{C}_+$. ■

34 <u>Comment</u>. Note that $\mathbb{C}_+$, i.e., the <u>closed</u> right-half of the complex plane, is the undesired set of the complex plane, where Σ_2 must not have any eigenvalue for Σ_2 to be exp. stable. Now for technical reasons (e.g., speed of response, damping, etc.) it is desirable to have a larger undesirable set, say $\mathcal{U} \subset \mathbb{C}$ as, for example, in Fig. 2.

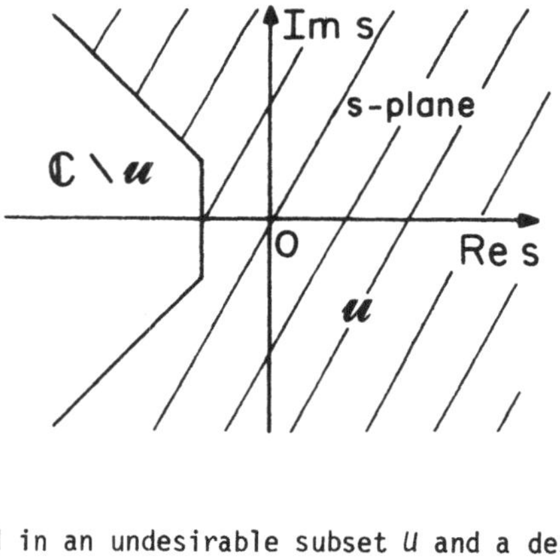

Fig. 2. $\mathbb{C}$ divided in an undesirable subset $\mathcal{U}$ and a desirable subset $\mathbb{C}\backslash\mathcal{U}$.

Hence the following stronger requirement for stability.

36 Definition. Consider feedback system Σ_2 shown in Fig. 1. Let $\mathcal{U}$ be a closed subset of $\mathbb{C}$ which is symmetric w.r.t. the real axis and such that $\mathcal{U} \supset \mathbb{C}_+$. We say that Σ_2 is $\mathcal{U}$-stable iff its PMD $\mathcal{D}_g$, described by (11)-(18), is well formed and its closed-loop eigenvalues are in $\mathbb{C}\backslash\mathcal{U}$.

In view of Fact 25 and Definitions 29, 30, and 36, we now have

37 Theorem [$\mathcal{U}$-stability of Σ_2]. If the Assumptions 1 hold, then

feedback system Σ_2, shown in Fig. 1, is $\mathcal{U}$-stable

iff

the char. poly. χ, given by (26), satisfies

38 $$Z[\chi] \subset \mathbb{C}\backslash\mathcal{U}.$$ ∎

Let us now consider the input-output properties of Σ_2.

Let

39 $$\mathcal{R}_{\mathcal{U}} := \{f \in \mathbb{R}_p(s) : f \text{ is analytic in } \mathcal{U}\} \subset \mathcal{R}(0)$$

denote the subring of exp. stable transfer functions (2.2.6), that are analytic in $\mathcal{U}$.

40 Corollary. Under the assumptions and condition of Theorem 37, feedback system Σ_2 has the property that all its closed-loop transfer functions have elements in $\mathcal{R}_{\mathcal{U}}$, more precisely the closed-loop transfer function

$$H : \begin{bmatrix} u_1 \\ u_2 \\ v_1 \\ d_0 \end{bmatrix} \longmapsto \begin{bmatrix} y_1 \\ y_2 \\ e_1 \\ e_2 \end{bmatrix} \in \mathcal{R}_{\mathcal{U}}^{4\times 4}.$$

41 Exercise. Prove Corollary 40.

42 Exercise. Consider the feedback system Σ_2. Let Assumptions 1 hold and let Σ_2 be $\mathcal{U}$-stable. Show that the $\mathcal{U}$-stability of Σ_2 still holds when the coefficients of the polynomials n_π, n_f, n_p, d_p, and d_c are subjected to

sufficiently small perturbations. (Hint: Note that $\mathcal{U}$ is closed.)

We shall now study conditions which ensure that the compensator $c = [\pi \mid f]$ is proper. Consider Fig. 1; let q denote closed-loop transfer function $q : u_1 \mapsto y_1$ and let $h_{y_2v_1}$ denote the system I/O map $h_{y_2v_1} : v_1 \mapsto y_2$. Straightforward calculations show that

45 $$q = f(1 + pf)^{-1} = d_p \chi^{-1} n_f$$

and

46 $$h_{y_2v_1} = p(1 + fp)^{-1}\pi = n_p \chi^{-1} n_\pi .$$

47 Theorem [Properness of f and π]. Consider system Σ_2 shown in Fig. 1. Let $p \in \mathbb{R}_{p,o}(s)$.
U.t.c.

(a)

48 $f \in \mathbb{R}_p(s)$ $(\mathbb{R}_{p,o}(s)$, resp.);

iff

49 $q \in \mathbb{R}_p(s)$ $(\mathbb{R}_{p,o}(s)$, resp.).

(b) Let $f \in \mathbb{R}_p(s)$; then

50 $\pi \in \mathbb{R}_p(s)$ $(\mathbb{R}_{p,o}(s)$, resp.);

iff

51 $p^{-1}h_{y_2v_1} \in \mathbb{R}_p(s)$ $(\mathbb{R}_{p,o}(s)$, resp.).

52 Comment. In view of (46), implication (50) ⇒ (51) is intuitively obvious: it says that if the compensator $c = [\pi \mid f]$ is proper, then the I/O map $h_{y_2v_1} \to 0$ as $|s| \to \infty$ at least as fast as the plant p. Furthermore, this is a necessary requirement on $h_{y_2v_1}$ in order to obtain a proper compensator.

53 Proof of Theorem 47

(a) ⇒ : Since p is strictly proper, if f is proper, then q is proper by (45).

(a) ⇐ : Equation (45) is equivalent to

54 $$f = q(1 - pq)^{-1};$$

hence, since p is strictly proper, q proper implies that f is proper.

(b) ⇒ : f proper and π proper imply that

$$(1 + fp)^{-1}\pi \in \mathbb{R}_p(s);$$

hence (51) follows from (46).

(b) ⇐ : By assumption p is strictly proper, f is proper, and (46) can be written as

55 $$(1 + fp)\, p^{-1} h_{y_2v_1} = \pi;$$

hence (51) ⇒ π is proper. ∎

From Corollary 40 and Theorem 47 we can state properties of the achievable I/O maps:

56 Corollary. Consider feedback system Σ_2, shown in Fig. 1, under the Assumptions 1 and let Σ_2 be $\mathcal{U}$-stable. Then

46 $$h_{y_2v_1} = p(1 + fp)^{-1}\pi = n_p\chi^{-1}n_\pi$$

satisfies:

57 $$p^{-1}h_{y_2v_1} \in \mathbb{R}_p(s),$$

58 $$\mathcal{P}[h_{y_2v_1}] \subset \mathbb{C}\backslash\mathcal{U},$$

59 $$Z[h_{y_2v_1}] \cap \mathcal{U} = \{Z[n_p] \cup Z[n_\pi]\} \cap \mathcal{U}.$$

In other words, any $\mathcal{U}$-zero of p and any $\mathcal{U}$-zero of π is necessarily a $\mathcal{U}$-zero of $h_{y_2v_1}$ because in (46) χ cannot cancel any $\mathcal{U}$-zero of p or π.

60 Exercise. For the system Σ_2, let Assumptions 1 hold and let Σ_2 be $\mathcal{U}$-stable. Prove the following:

(a) Let $H_{yu} : (u_1, u_2, v_1) \mapsto (y_1, y_2)$, then, with q defined by (45),

$$H_{yu} = \left[\begin{array}{c|c|c} f(1+pf)^{-1} & -fp(1+fp)^{-1} & (1+fp)^{-1}\pi \\ \hline pf(1+pf)^{-1} & p(1+fp)^{-1} & p(1+fp)^{-1}\pi \end{array}\right],$$

$$= \left[\begin{array}{c|c|c} q & -qp & (1-qp)\pi \\ \hline pq & p(1-qp) & p(1-qp)\pi \end{array}\right].$$

(b) If $H_{yu} \in \mathcal{R}_{\mathcal{U}}^{2\times 3}$, then all transfer functions from any (u_1, u_2, v_1, d_0) to any $(y_1, y_2, e_1, e_2, y_p)$ are in $\mathcal{R}_{\mathcal{U}}$.

(c) Assume, in addition, that $p \in \mathcal{R}_{\mathcal{U}} \cap \mathbb{R}_{p,o}(s)$; then the I/O map $h_{y_2v_1}$ can be realized by a $\mathcal{U}$-stable feedback system Σ_2 with π and $f \in \mathbb{R}_p(s)$ if and only if

$$h_{y_2v_1} = p(1-qp)\pi$$

for some $q \in \mathcal{R}_{\mathcal{U}}$ and $\pi \in \mathbb{R}_p(s)$ s.t. $(1-qp)\pi \in \mathcal{R}_{\mathcal{U}}$. ✱

We shall now study robust asymptotic tracking throughout, the inputs v_1 to be tracked (specified in terms of Laplace transforms) are in the class

61 $$\Psi := \{\hat{v}_1 = \psi^{-1}\mu : \psi, \mu \in \mathbb{R}[s] \text{ with } \partial[\mu] < \partial[\psi]\},$$

where ψ is a given monic polynomial and μ is arbitrary; furthermore,

62 $$Z[\psi] \subset \mathbb{C}_+ \subset \mathcal{U}$$

and

63 $$Z[\psi] \cap Z[n_p] = \phi.$$

The arbitrariness of μ in (61) corresponds to selecting arbitrary initial conditions in the dynamical system generating all the reference inputs v_1 in the class Ψ.

We will see later that (63) is a necessary and sufficient condition that the plant $p \in \mathbb{R}_{p,o}(s)$ must satisfy in order that the system of Fig. 1 with a proper compensator $c = [\pi \,\vdots\, f]$ can track every input in the class Ψ. (This is a well-known fact for the multivariable case; see, e.g. [Dav.1], [Des.2].)

Let

64[a] $$\eta(t) := -(y_2 - v_1)(t)$$

be the tracking error (see Fig. 1), whence in terms of the Laplace transform

64[b] $$\hat{\eta} = -(h_{y_2v_1} - 1)\hat{v}_1 .$$

65 Definition. We say that feedback system Σ_2, shown in Fig. 1, achieves robust asymptotic tracking over the class Ψ if and only if the following three conditions are satisfied:

(1) system Σ_2, shown in Fig. 1, is U-stable;

66 (2) for any input $v_1 \in \Psi$, the tracking error $\eta(t) \to 0$ exponentially as $t \to \infty$;

(3) the tracking requirement (66) holds for any perturbed plant $\tilde{p} = \tilde{n}_p \tilde{d}_p^{-1} \in \mathbb{R}_{p,o}(s)$ where the polynomials $\tilde{n}_p$ and $\tilde{d}_p$ are arbitrary subject to

(i) $\tilde{n}_p$ and $\tilde{d}_p$ are coprime;

(ii) $Z[\psi] \cap Z[\tilde{n}_p] = \phi$;

(iii) the perturbed closed-loop system Σ_2 is exp. stable (i.e. $Z[\tilde{\chi}] \subset \mathring{\mathbb{C}}_-$ where $\tilde{\chi} := \tilde{d}_p d_c + \tilde{n}_p n_f$). ∎

Now, for any input $v_1 \in \Psi$, (64[b]) and (46) give

70 $$\hat{\eta} = \frac{[n_p(n_\pi - n_f) - d_p d_c]}{\chi} \frac{\mu}{\psi}$$

From (70), we obtain the last theorem of this section.

71 Theorem [Robust asymptotic tracking]. Consider feedback system Σ_2, shown in Fig. 1, under the Assumptions 1. Let the class Ψ of inputs v_1 to be tracked be specified by (61), (62), (63).

U.t.c.

72 feedback system Σ_2 achieves robust asymptotic tracking over the class Ψ

iff

73 (i) $Z[\chi] \subset \mathbb{C}\backslash\mathcal{U}$;

74 (ii) $\psi | d_c$;

75 (iii) $\psi | (n_\pi - n_f)$.

76 Comments. (a) Since $\chi = d_p d_c + n_p n_f$, (73) and (74) imply that

63 $$Z[\psi] \cap Z[n_p] = \phi.$$

Algorithm (5.3.55) below shows the converse; namely, if (63) holds, then there exists a proper f and π such that robust asymptotic tracking over the class Ψ is achieved .

(b) From (70), conditions (73), (74), and (75) imply that $\forall \zeta \in Z[\psi]$, $h_{y_2 v_1}(\zeta) = 1$; and, if ζ is a kth order zero of ψ, then $h^{(i)}_{y_2 v_1}(\zeta) = 0$ for $i = 1, 2, \cdots, k - 1$. This is as expected from Exercise 3.3.2.28.

78 Proof of Theorem 71. $\Leftarrow$. From (73)-(75) and Corollary 40, it follows that $\hat{\eta}$, given by (70), is strictly proper and analytic in $\mathbb{C}_+$; hence (66) follows. Furthermore, this conclusion still holds for any $\tilde{n}_p$, $\tilde{d}_p$ as long as $Z[\tilde{\chi}] \subset \mathring{\mathbb{C}}_-$, as required by the definition of robust asymptotic tracking.

$\Rightarrow$. Robust asymptotic tracking implies (73)-(75): indeed, (73) is required for $\mathcal{U}$-stability (Theorem 37), and the robust asymptotic tracking condition (66), for any arbitrary $\tilde{n}_p$, $\tilde{d}_p$ as in Definition 65, requires ψ to cancel in (70) for any such $\tilde{n}_p$, $\tilde{d}_p$. Hence (74)-(75) must hold, i.e., $\psi | d_c$ and $\psi | n_\pi - n_f$

More precisely, consider the case where ψ has a simple zero at $\zeta \in \mathbb{C}_+$. Since robust asymptotic tracking is achieved, from (70) we have

79 $$n_p(\zeta)\big(n_\pi(\zeta) - n_f(\zeta)\big) - d_p(\zeta)d_c(\zeta) = 0$$

and (79) holds when arbitrary small perturbations are applied to the coefficients of n_p and d_p. Consider two cases:

Case 1. $d_p(\zeta) \neq 0$. Consider the following perturbation in n_p and d_p: for some sufficiently small $\varepsilon > 0$, multiply all the coefficients of d_p by $(1 - \varepsilon)$; then, by robust asymptotic tracking, we have

80 $$n_p(\zeta)\big(n_\pi(\zeta) - n_f(\zeta)\big) - (1 - \varepsilon)d_p(\zeta)d_c(\zeta) = 0.$$

By assumption $d_p(\zeta) \neq 0$, and, by (63), $n_p(\zeta) \neq 0$; hence equations (79) and (80) imply that

81 $$n_\pi(\zeta) - n_f(\zeta) = d_c(\zeta) = 0.$$

Case 2. $d_p(\zeta) = 0$. By assumption, $\partial[d_p] \geq 1$, so an arbitrarily small perturbation of a suitable coefficient of d_p yields a new polynomial $\tilde{d}_p$ such that $\tilde{d}_p(\zeta) \neq 0$; robust asymptotic tracking then implies that

82 $$n_p(\zeta)\Big(n_\pi(\zeta) - n_f(\zeta)\Big) - \tilde{d}_p(\zeta) d_c(\zeta) = 0.$$

Since $n_p(\zeta) \neq 0$, $\tilde{d}_p(\zeta) \neq 0$, and $d_p(\zeta) = 0$, (79) and (82) imply (81).

Therefore, we have proven that robust asymptotic tracking implies (74) and (75), when ψ has simple zeros. When ψ has multiple zeros, a similar proof can be constructed. ∎

5.3. Design

We propose to use the theorems of Sec. 5.2 to show how one may obtain a compensator for system Σ_2, shown in Fig. 5.2.1, to achieve a prescribed I/O map $h_{y_2v_1}$ and to achieve robust asymptotic tracking over a given class Ψ.

8 Data. We are given

9 (1) the closed set $\mathcal{U} \supset \mathbb{C}_+$ of undesirable closed-loop eigenvalue locations

(2) the plant

10 $$p = \frac{n_p}{d_p} = \frac{n_{pu}\, n_{ps}}{d_p} \in \mathbb{R}_{p,o}(s),$$

where

11 (i) n_p and d_p are coprime,

12 (ii) $n_p = n_{pu}\, n_{ps}$ with $Z[n_{pu}] \subset \mathcal{U}$, $Z[n_{ps}] \subset \mathbb{C}\backslash\mathcal{U}$;

(3) the specified I/O map $h_{y_2v_1}$ is given by

13 $$h_{y_2v_1} = \frac{n_{pu}\, n_{h1}}{d_h}$$

where

14 (i) $p^{-1} h_{y_2 v_1} \in \mathbb{R}_p(s)$ (so that by Theorem 5.2.47 the compensator will be proper);

15 (ii) (n_{h_1}, d_h) are coprime polynomials;

16 (iii) $Z[d_h] \subset \mathbb{C} \backslash \mathcal{U}$ (a consequence of the required $\mathcal{U}$-stability of the system).

20 Comments. (a) Recall that, from (5.2.26), $\chi = d_c d_p + n_f n_p$ and that by (5.2.45) and (5.2.46),

21 $$q = d_p \, \chi^{-1} \, n_f,$$

22 $$h_{y_2 v_1} = n_p \, \chi^{-1} \, n_\pi .$$

(b) By Theorem 5.2.47 and (14) once q in (21) is proper, the compensator $c = [\pi \,\vdots\, f]$ is proper, and then by Theorem 5.2.37 and (10), the $\mathcal{U}$-stability of feedback system Σ_2 is equivalent to $Z[\chi] \subset \mathbb{C} \backslash \mathcal{U}$. Therefore, in (22), any factor common to n_p and χ must have all its zeros in $\mathbb{C} \backslash \mathcal{U}$. Hence n_{pu} must be a factor of the numerator of $h_{y_2 v_1}$ as indicated in (13).

(c) Equating the two expressions (13) and (22) for $h_{y_2 v_1}$, and canceling n_{pu}, we get

23 $$n_\pi = \frac{n_{h_1} \, \chi}{d_h \, n_{ps}} .$$

Note that the $h_{y_2 v_1}$ specified by (13) will be achieved by the closed-loop system of Fig. 1 if and only if (23) holds.

(d) The algorithm below actually establishes the converse of Corollary 5.2.56: indeed, conditions (13)-(16) are identical with (5.2.46), (5.2.57)-(5.2.59).

24 Algorithm [For Σ_2 to achieve the prescribed I/O map $h_{y_2 v_1}$].

Step 1. Choose $\chi \in \mathbb{R}[s]$ s.t.

25 (i) $d_h \, n_{ps} \mid n_{h_1} \chi$ (hence (23) will give n_π as a polynomial);

26 (ii) $Z[\chi] \subset \mathbb{C} \backslash \mathcal{U}$ (to achieve $\mathcal{U}$-stability);

27 (iii) $\partial\chi \geq 2\,\partial d_p - 1$. (See Comment (35)(b), below.)

<u>Step 2</u>. Choose $n_f \in \mathbb{R}[s]$ s.t.

30 (i) $\partial n_f \leq \partial\chi - \partial d_p$ (equiv. q, in (21), is <u>proper</u>);

31 (ii) $d_p \mid (\chi - n_p n_f)$

and set

32
$$d_c := \frac{\chi - n_p n_f}{d_p}$$

33
$$n_\pi := \frac{n_{h_1}\ \chi}{d_h\ n_{ps}}\ .$$

End of Algo

35 <u>Comments</u>. (a) The first step is to choose the closed-loop characteristic polynomial. Note that the algorithm <u>does not</u> guarantee that n_f and d_c are coprime, nor that $[n_\pi \,\vdots\, n_f]$ and d_c are coprime. Examples show that complete cancellation may occur! Note that in any case, the analysis in (5.2.10) and the subsequent Definition 5.2.29 establish that $d_c d_p + n_f n_p$ is the characteristic polynomial of (the PMD $\mathcal{D}_g$ of) Σ_2. Of course, cancellation of common factors in $[n_\pi \,\vdots\, n_f]$ and d_c will simplify the realization of the compensator and the resulting closed-loop system will still have all its closed-loop eigenvalues in $\mathbb{C}\backslash\mathcal{U}$ (see (26)).
(b) Note that (31)-(32) is equivalent to the solution of the equation

36
$$n_f n_p + d_c d_p = \chi,$$

where n_p, d_p, and χ are given and n_f, d_c are unknown. Since (n_p, d_p) is coprime, the Bezout identity states that there are polynomials u, v (obtained by the Euclidean algorithm [MacL.1, p. 122]) such that

37
$$u n_p + v d_p = 1.$$

Hence, all polynomial solutions (n_f, d_c) of (36) are given by

38
$$\begin{aligned} n_f &= u\chi - d_p m \\ d_c &= v\chi + n_p m \end{aligned} \qquad \text{for any } m \in \mathbb{R}[s].$$

(To see this: multiply (37) by χ to obtain a particular solution $(u\chi, v\chi)$ of (36) and add to it the general solution $(-d_p m, n_p m)$ of the homogeneous equation corresponding to (36); see also Theorem 6.2.39 below.) Conditions under which (38) yields a proper feedback compensator $f = n_f/d_c$ are easily obtained: by (21) and Theorem 5.2.47 we have

30 $$\partial[n_f] \le \partial[\chi] - \partial[d_p]$$

$\Leftrightarrow$ q proper $\Leftrightarrow$ f proper.

Hence by Theorem 5.2.47, (14) and (30) are necessary and sufficient conditions in order that the compensator $c = [\pi \,\vdots\, f]$ be proper. Moreover, under these conditions and since $p \in \mathbb{R}_{p,o}(s)$ it follows from (36) that

39 $$\partial[d_c] = \partial[\chi] - \partial[d_p].$$

Observe now that the least degree solution n_f of (36) is given by (38), by dividing $u\chi$ by d_p (note that m and n_f are then resp. quotient and remainder). Hence for that solution

40 $$\partial[n_f] \le \partial[d_p] - 1.$$

So if we apply condition (27), equiv.

41 $$\partial[d_p] - 1 \le \partial[\chi] - \partial[d_p],$$

then we have that the least degree solution n_f satisfies (30). Hence (27) is a sufficient condition for the existence of a proper feedback compensator f characterized by (38) and (30), or equiv. (30)-(32), and the precompensator π is then proper by (14) and Theorem 5.2.47.

(c) In (31) χ and n_p are fixed; hence we have to adjust the coefficients of $n_f(s) := \sum_{j=0}^{n} n_{f,j}s^j$ so that $\chi - n_p n_f$ is a multiple of d_p.

For simplicity suppose that d_p has $m := \partial[d_p]$ distinct zeros $(p_k)_{k=1}^{m}$; then (31) is equivalent to

42 $$\chi(p_k) - n_p(p_k)n_f(p_k) = 0 \quad \forall k = 1, 2, \cdots, m = \partial[d_p].$$

Since χ and n_p are known polynomials, with $n_p(p_k) \neq 0$ $\forall k$ (because (n_p, d_p) is coprime), (31) $\Leftrightarrow$ (42) is a system of m linear equations in n + 1 unknowns $(n_{f,j})_o^n \in \mathbb{R}^{n+1}$; moreover, if $n \geq m - 1$, the system (42) has a matrix with a nonsingular $m \times m$ Vandermonde submatrix, (since $p_k \neq p_\ell$, $\forall k \neq \ell$). Hence for $n \geq m - 1$, (42) has (a) solution(s) $(n_{f,j})_o^n$ and n_f can be computed in this manner; moreover, the solution (40) is obtained if we set $n = m - 1 = \partial[d_p] - 1$.

Summary. Algorithm 24 delivers the specified I/O map $h_{y_2v_1}$: the design gives a proper $f = n_f/d_c$, a proper $\pi = n_\pi/d_c$, and a $\mathcal{U}$-stable char. poly. χ.

45 Example. Given $p(s) = (s + 1)/(s - 1)^3$. To achieve the I/O map $h_{y_2v_1} = (s + 1)^{-2}$, we apply Algorithm 24: let $\chi(s) = (s + 1)^5$; then $n_\pi(s) = (s + 1)^2$, $n_f(s) = 8(3s^2 - 2s + 1)$ and $d_c(s) = (s + 1)(s + 7)$.

The design of compensators for robust asymptotic tracking is based on Theorem 5.2.71. It is best explained in terms of an algorithm-

55 Algorithm [Robust asymptotic tracking]

Data:

56 (1) $\mathcal{U}$ and p are specified as in (9)-(12),

57 (2) $\psi \in \mathbb{R}[s]$ with $Z[\psi] \subset \mathbb{C}_+$ and

58 $$Z[\psi] \cap Z[n_p] = \phi.$$

Step 1. Choose $\chi \in \mathbb{R}[s]$ s.t.

61 $Z[\chi] \subset \mathbb{C}\backslash\mathcal{U}$,

62 $\partial[\chi] \geq \partial\psi + 2\partial d_p - 1$

Step 2. Choose $n_f \in \mathbb{R}[s]$ s.t.

65 $\partial n_f \leq \partial\chi - \partial d_p$,

66 $(\psi d_p) | (\chi - n_f n_p)$.

Set

67 $$d_c := \frac{\chi - n_f n_p}{d_p}\ .$$

<u>Step 3</u>. Choose $n_\pi \in \mathbb{R}[s]$ s.t.

70 $\partial n_\pi \le \partial\chi - \partial d_p$,

71 $\psi | (n_\pi - n_f)$.

<u>Justification</u>

By (67), $\chi = d_c d_p + n_f n_p$ is the char. poly. of the closed-loop system. Robust asymptotic tracking is achieved: indeed, recall Theorem 5.2.71: note (61); by (66) and (67), $\psi | d_c$; by (71) $\psi | (n_\pi - n_f)$.

Now recall Theorem 5.2.47: in view of (21), (65) implies that q is proper, hence f is proper. Hence by (67) $\partial[d_c] = \partial[\chi] - \partial[d_p]$, s.t. (70) implies that π is proper.

Note that by (62) and (65), $n_f = \sum_{j=0}^{n} n_{f,j} s^j$ has a coefficient vector $(n_{f,j})_0^n$ s.t. n may be chosen s.t. $n = \partial\psi + \partial d_p - 1$; therefore, there are enough parameters to satisfy (66). ∎

Chapter 6. The Closed-Loop Eigenvalue Placement Problem

6.1. Introduction

In this chapter we are given a strictly proper multi-input multi-output plant P and we wish to design a unity feedback system Σ using a compensator C. This compensator C should achieve a given closed-loop eigenvalue placement, which is specified in terms of a nonsingular polynomial matrix D_k resulting in the closed-loop characteristic polynomial $\chi(s) = \det D_k(s)$. D_k specifies completely the zero-input pseudo-state trajectories of feedback system Σ (Fact 6.2.30). The compensator, being a polynomial matrix fraction, has a numerator and denominator that satisfy compensator equation (6.2.32). Theorem 6.2.39 describes the class of polynomial matrix solutions of this equation. Theorem 6.2.61 characterizes all internally proper solutions. Finally, Theorem 6.2.84 gives sufficient conditions for the existence of internally proper solutions.

6.2. The Compensator Problem

1 Problem COMP. We are given a plant P s.t.

2 $P \in \mathbb{R}_{p,o}(s)^{n_o \times n_i}$ has an int. pr. r.c.f. (N_{pr}, D_{pr}) s.t.

3 D_{pr} is column-reduced with column degrees k_j, $j \in \underline{n}_i$,
and

4 with highest column degree coefficient matrix $D_{ph} = I_{n_i}$.

We must find a compensator C s.t.

5 $C \in \mathbb{R}_p(s)^{n_i \times n_o}$ has an int. pr. ℓ.f. $(D_{c\ell}, N_{c\ell})$
with

6 $D_{c\ell}$ row-reduced,
s.t.

7 the feedback system Σ, shown in Fig. 4.3.1, is exp. stable with prescribed closed-loop characteristic polynomial $\chi(s)$.

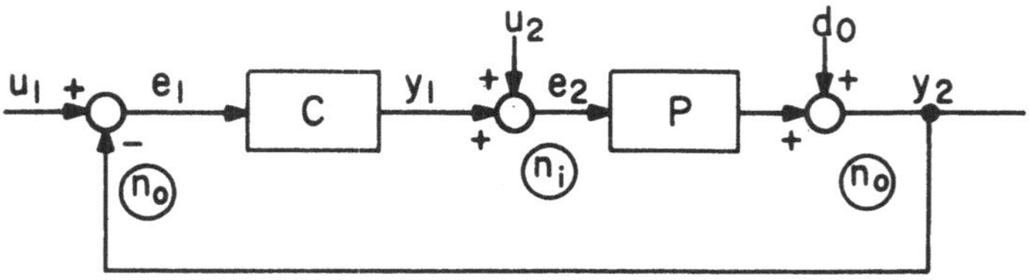

Fig. 4.3.1. The feedback system Σ under consideration.

8 Comments. (a) An int. pr. r.c.f. of P as in (2), (3) can always be obtained from any r.f. of P by using suitable e.c.o.'s to extract a g.c.r.d. and make D_{pr} column-reduced.

(b) Assumption (3) is equivalent to

9 $$D_{pr} = D_{p-} \operatorname{diag}[s^{k_j}]_{j=1}^{n_i},$$

where D_{p-} is biproper and

10 $$D_{p-}(\infty) = D_{ph}.$$

(c) In order to achieve $D_{ph} = I$ as in Assumption (4), a change of plant-input-coordinates may be necessary. The plant gives

11 $$y_2 = Pe_2 = N_{pr} D_{pr}^{-1} e_2.$$

If $D_{ph} \neq I$, we set

12 $$\bar{e}_2 = D_{ph}^{-1} e_2;$$

hence

13 $$y_2 = \bar{P}\,\bar{e}_2, \quad \text{where} \quad \bar{P} = N_{pr}(D_{ph}^{-1} D_{pr})^{-1}.$$

(d) By Corollary 4.3.26, the feedback Σ obeys assumptions IS, WP, and WF of Sec. 4.2; furthermore, Σ is exp. stable if and only if $Z[\chi] \subset \mathring{\mathbb{C}}_-$, where $\chi(s)$ equals the RHS of (4.3.8). ∎

16 Analysis. Analyzing Σ, as in Sec. 4.3, we obtain

17 $$\chi(s) = \det[D_{c\ell}D_{pr} + N_{c\ell}N_{pr}](s),$$

18 $$H_{y_2u_1} = N_{pr}[D_{c\ell}D_{pr} + N_{c\ell}N_{pr}]^{-1}N_{c\ell}.$$

Thus setting

19 $$D_k := D_{c\ell}D_{pr} + N_{c\ell}N_{pr},$$

we have

20 $$\chi(s) = \det D_k(s).$$

Let us view P and C as specified by the PMD's $\mathcal{D}_p = [D_{pr}, I, N_{pr}, 0]$ and $\mathcal{D}_c = [D_{c\ell}, N_{c\ell}, I, 0]$, resp.. Call ξ_p, ξ_c, resp., the corresponding pseudo-states. As in Secs. 4.2 and 4.3, we consider the PMD $\mathcal{D}_{yu} = [D_g, N_\ell, N_r, 0]$ where $\xi := (\xi_c^T, \xi_p^T)^T$ and

25 $$N_r = \text{diag}[I \mid N_{pr}],$$

26 $$N_\ell = \text{diag}[N_{c\ell} \mid I],$$

27 $$D_g = \left[\begin{array}{c|c} D_{c\ell} & N_{c\ell}N_{pr} \\ \hline -I & D_{pr} \end{array}\right].$$

For zero input, the differential equation representing the system reads (after performing on (27) the e.r.o. $\rho_1 \leftarrow \rho_1 + D_{c\ell}\rho_2$),

28 $$\begin{bmatrix} 0 & D_k \\ -I & D_{pr} \end{bmatrix}\begin{bmatrix} \xi_c(t) \\ \xi_p(t) \end{bmatrix} = \begin{bmatrix} \theta \\ \theta \end{bmatrix}$$

or equivalently,

29 $$\begin{cases} D_k(p)\xi_p(t) = \theta, \\ \xi_c(t) = D_{pr}(p)\xi_p(t). \end{cases}$$

From (20), (25), (26), and (29), we deduce the following results:

30 Fact. The polynomial matrix $D_k = D_{c\ell}D_{pr} + N_{c\ell}N_{pr}$ specifies completely the z-i p.s trajectories $\xi(\cdot)$ of the feedback system Σ. By (29), D_k controls directly the dynamics of the z-i p.s. trajectories $\xi_p(\cdot)$ of the plant. By (20) the determinant of D_k is the char. pol. of Σ. ∎

31 Comment. If D_k is chosen to be diagonal, then each diagonal entry specifies independently the characteristic polynomial of each entry of $\xi_p(\cdot)$

Solving for the Compensator. In design, N_{pr} and D_{pr} are given and they satisfy (1)-(4). The matrix D_k is chosen so that $Z[\det D_k(s)] \subset \mathbb{C}_-$, for exp. stability, and so that the compensator equation

32 $$X\,D_{pr} + Y\,N_{pr} = D_k$$

has (a) solution(s)

33 $$(X, Y) \in \mathbb{R}[s]^{n_i \times n_i} \times \mathbb{R}[s]^{n_i \times n_o},$$

s.t.

34 the ℓ.f. (X, Y) is int. pr. and X is row reduced. ∎

Note now that, since by (2)-(4), (N_{pr}, D_{pr}) is a r.c.f. of P, Theorem 2.4.1.25.R delivers a generalized Bezout identity: there exist six polynomial matrices

35 $$U_{pr}, V_{pr}, D_{p\ell}, N_{p\ell}, U_{p\ell}, V_{p\ell} \in E(\mathbb{R}[s]),$$

s.t.

36 $$\begin{array}{c} \\ n_i \\ n_o \end{array}\!\!\begin{array}{c} \begin{array}{cc} n_i & n_o \end{array} \\ \left[\begin{array}{c|c} V_{pr} & U_{pr} \\ \hline -N_{p\ell} & D_{p\ell} \end{array}\right] \end{array}\begin{array}{c} \begin{array}{cc} n_i & n_o \end{array} \\ \left[\begin{array}{c|c} D_{pr} & -U_{p\ell} \\ \hline N_{pr} & V_{p\ell} \end{array}\right] \end{array} = \left[\begin{array}{c|c} I_{n_i} & 0 \\ \hline 0 & I_{n_o} \end{array}\right].$$

Moreover,

37 $$(D_{p\ell}, N_{p\ell}) \text{ is a } \ell\text{.c.f. of } P$$

and without loss of generality

38 $D_{p\ell}$ is row-reduced.

39 Theorem [Polynomial solutions of comp. eq. (32)]. Consider (32), where $P = N_{pr}D_{pr}^{-1}$ satisfies (2)-(4). U.t.c.

40 $$(X, Y) \in \mathbb{R}[s]^{n_i \times n_i} \times \mathbb{R}[s]^{n_i \times n_o} \text{ is a solution of (32)}$$

$\Leftrightarrow$

$$\exists\, N_k \in \mathbb{R}[s]^{n_i \times n_i} \quad \text{s.t.}$$

41 $$X = D_k V_{pr} - N_k N_{p\ell},$$

42 $$Y = D_k U_{pr} + N_k D_{p\ell},$$

where U_{pr}, V_{pr}, $D_{p\ell}$, $N_{p\ell}$ are elements of the generalized Bezout identity (36). Moreover,

43 $$(X, Y) \text{ is } \ell.\text{c.} \Leftrightarrow (D_k, N_k) \text{ is } \ell.\text{c.}$$ ∎

44 Comments. (a) Given a solution (X, Y), we set $(X, Y) = (D_{c\ell}, N_{c\ell})$, i.e., we obtain the compensator $C = D_{c\ell}^{-1}N_{c\ell} = X^{-1}Y$, provided that det $X \neq 0$. The additional requirement that (X, Y) be an int. pr. ℓ.f. with X row-reduced is investigated below.

(b) Equations (41) and (42) constitute a global parametrization by the polynomial matrix N_k of all the solutions of (32): N_k determines (X, Y) uniquely, and vice versa. Note that N_k may be viewed as the quotient of the division of $D_k U_{pr}$ on the right by $D_{p\ell}$: Write (42) as

45 $$D_k U_{pr} = -N_k D_{p\ell} + Y \quad \text{s.t.} \quad YD_{p\ell}^{-1} \in \mathbb{R}_{p,o}(s)^{n_i \times n_o}.$$

The quotient N_k then determines X by (41).

46 Proof of Theorem 39. (a) Proof of $\Leftarrow$: By assumption (41) and (42) hold. For any N_k substitute X and Y given by (41) and (42) into (32); for the LHS of (32) we obtain

47 $$D_k(V_{pr}D_{pr} + U_{pr}N_{pr}) + N_k(-N_{p\ell}D_{pr} + D_{p\ell}N_{pr}).$$

Now considering the (1, 1) entry and the (2, 1) entry of (36), (47) reduces to D_k, as required by (32). Hence $\forall N_k$, X and Y, defined by (41) and (42), are solutions of (32).

(b) Proof of $\Rightarrow$: By assumption (X, Y) is a solution of (32). First, consider the (1, 1) entry of (36) and multiply on the left by D_k; then

$$D_k V_{pr} D_{pr} + D_k U_{pr} N_{pr} = D_k .$$

Hence

48 $$X_p := D_k V_{pr}, \; Y_p := D_k U_{pr}$$

specifies a particular solution of (32).

Second, since (32) is a linear equation we seek the general solution of the homogeneous equation

49 $$X D_{pr} + Y N_{pr} = 0.$$

Recall that the (2, 1) entry of (36) gives $-N_{p\ell} D_{pr} + D_{p\ell} N_{pr} = 0$. For any polynomial solution (X_h, Y_h) of the homogeneous equation (49) let

50 $$N_k := Y_h D_{p\ell}^{-1};$$

then, since (X_h, Y_h) is a solution of (49),

$$X_h = -Y_h N_{pr} D_{pr}^{-1} = -Y_h D_{p\ell}^{-1} N_{p\ell} = -N_k N_{p\ell} .$$

Hence any solution (X_h, Y_h) of (49) is of the form

51 $$X_h = -N_k N_{p\ell}, \qquad Y_h = N_k D_{p\ell} .$$

Finally, we claim that N_k, defined by (50), is a polynomial matrix; indeed, using (36),

$$N_k = N_k (N_{p\ell} U_{p\ell} + D_{p\ell} V_{p\ell}) = -X_h U_{p\ell} + Y_h V_{p\ell} \in E(\mathbb{R}[s]).$$

Hence we have exhibited a particular solution (X_p, Y_p) in (48) and we have shown that <u>any</u> solution (X_h, Y_h) is of the form (51), where N_k is a <u>polynomial</u> matrix; consequently any solution (X, Y) of (32) is of the form (41) and (42).

(c) From (41) and (42) we have

52 $$[X \mid Y] = [D_k \mid N_k]\left[\begin{array}{c|c} V_{pr} & U_{pr} \\ \hline -N_{p\ell} & D_{p\ell} \end{array}\right].$$

The last matrix in the RHS is unimodular (see (36)); hence $\mathrm{rk}[X \mid Y](s) = \mathrm{rk}[D_k \mid N_k](s)$, $\forall s \in \mathbb{C}$ and (43) follows. ∎

53 <u>Exercise.</u> Prove the following. If in the analysis (16) we had used the characteristic polynomial (4.3.9), we would have obtained the comp. eqn. $D_{p\ell}X + N_{p\ell}Y = D_k$. All polynomial solutions of this equation are of the form

$$X_r = V_{p\ell}D_k - N_{pr}N_k$$

$$Y_r = U_{p\ell}D_k + D_{pr}N_k.$$ ∎

We now give conditions under which the solutions (X, Y) of the comp. equ. (32) are s.t. $X^{-1}Y$ is an int. proper left fraction, [Emr.1].

61 <u>Theorem</u> [Internally proper solutions of comp. equ. (32)]. Consider the comp. equ. (32), where the plant P is given by

2 $P \in \mathbb{R}_{p,o}(s)^{n_o \times n_i}$ has an int. pr. r.c.f. (N_{pr}, D_{pr})

s.t.

3 D_{pr} is column reduced with column degrees k_j, $j \in \underline{n}_i$

and

4 with highest column degree coefficients matrix $D_{ph} = I_{n_i}$.

U.t.c.

the compensator equation, given by

32 $$XD_{pr} + YN_{pr} = D_k,$$

has a solution (X, Y) s.t.

33 $$(X, Y) \in \mathbb{R}[s]^{n_i \times n_i} \times \mathbb{R}[s]^{n_i \times n_o}$$

and

the left fraction (X, Y) is int. pr. with

62 X row-reduced with row degrees r_i, $i \in \underline{n}_i$

and highest row-degree coefficient matrix X_h

if and only if

(a)

63 D_k is row-column-reduced with row and column powers r_i, $i \in \underline{n}_i$, resp. k_j, $j \in \underline{n}_i$,

or equiv. (see criterion (3.3.1.63)),

64 $$D_k(s) = \text{diag}[s^{r_i}]_{i=1}^{n_i} \, D_{k-}(s) \, \text{diag}[s^{k_j}]_{j=1}^{n_i},$$

where $D_{k-} \in \mathbb{R}(s)^{n_i \times n_i}$ is <u>biproper</u> and the highest degree coefficient matrix D_{kh} of D_k satisfies

65 $$D_{kh} = D_{k-}(\infty);$$

(b)

66 for the given D_k, (X,Y) is a polynomial matrix solution of comp. eqn. (32) s.t. the row degrees of Y satisfy

$$\partial_{r_i}[Y] \leq r_i \quad \forall i \in \underline{n}_i.$$

Moreover, under conditions (a)-(b) we have that the highest degree coefficient matrices of D_{pr}, X, and D_k are related by

67 $$D_{kh} = X_h \, D_{ph}$$

s.t., if $D_{ph} = I$ and $D_{kh} = I$, then $X_h = I$. ∎

67 <u>Comment</u>. From Comments (44) and statement (62), which is an expansion of (34), the order of the solving compensator C is $\sum_{i=1}^{n_i} r_i$, where the r_i's are the row powers of the chosen matrix D_k (see (63)). We shall see below that <u>if these row powers</u> r_i <u>are sufficiently large</u>, then solutions (X, Y) satisfying (66) will exist. Hence we must choose $\chi(s) = \det D_k(s)$, the characteristic polynomial of Σ, to be of sufficiently large degree for problem COMP to have a solution.

70 <u>Proof of Theorem 61</u>. $\Rightarrow$: Since $P \in \mathbb{R}_{p,o}(s)^{n_o \times n_i}$ has an int. pr. r.c.f. (N_{pr}, D_{pr}), with D_{pr} column-reduced, we have

71 $D_{pr} = D_{p-}\ \mathrm{diag}[s^{k_j}]$ and $N_{pr}\ \mathrm{diag}[s^{-k_j}] \in \mathbb{R}_{p,o}(s)^{n_o \times n_i}$,

where $D_{p-} \in \mathbb{R}(s)^{n_i \times n_i}$ is biproper and in (4)

72 $$D_{ph} = D_{p-}(\infty).$$

Since $X^{-1}\ Y \in \mathbb{R}_p(s)^{n_i \times n_o}$ has an int. pr. ℓ.f. (X, Y) with X row-reduced, we have similarly

73 $X = \mathrm{diag}[s^{r_i}]\ X_-$ and $\mathrm{diag}[s^{-r_i}]\ Y \in \mathbb{R}_p(s)^{n_i \times n_o}$,

where $X_- \in \mathbb{R}(s)^{n_i \times n_i}$ is biproper and in (62)

74 $$X_h = X_-(\infty).$$

Hence, by (71)-(74), the comp. eqn. (32) reads

75 $$D_k = \mathrm{diag}[s^{r_i}]\ D_{k-}\ \mathrm{diag}[s^{k_j}],$$

where

76 $D_{k-} := X_-\ D_{p-} + \mathrm{diag}[s^{-r_i}]\ YN_{pr}\ \mathrm{diag}[s^{-k_j}]$ is biproper

with

77 $$D_{kh} := D_{k-}(\infty) = X_-(\infty)\ D_{p-}(\infty) =: X_h\ D_{ph}.$$

Hence (64) follows from (75)-(76), and (66) follows from (73^b). Note also that (77) is (67).

⇐ : We note again that, by the plant assumptions, (71)-(72) hold; moreover, (66) implies (73^b), where the r_j are the row powers of D_k in (64). Using now comp. eqn. (32), we have successively

$$\mathrm{diag}[s^{r_i}]\ D_{k-}\ \mathrm{diag}[s^{k_j}] = XD_{p-}\ \mathrm{diag}[s^{k_j}] + Y\ N_{pr},$$

and

78 $$\mathrm{diag}[s^{-r_i}]\ X = \left\{D_{k-} - (\mathrm{diag}[s^{-r_i}]\ Y)(N_{pr}\ \mathrm{diag}[s^{-k_j}])\right\}\ D_{p-}^{-1}.$$

Note that on RHS of (78) we obtain a product of two biproper factors, the first one being biproper becasue ($\text{Diag}[s^{-r_i}]\, Y$) is proper and ($N_{pr}\, \text{diag}[s^{-k_j}]$) is strictly proper by (73^b) and (71^b). Hence

79 $$X = \text{diag}[s^{r_i}]\, X_- \text{ with } X_- \text{ biproper.}$$

Therefore, X is row-reduced with row degrees r_i and highest row-degree coefficient matrix

$$X_h = X_-(\infty).$$

Moreover, since $\text{diag}[s^{-r_i}]\, Y$ is proper by (73^b), it follows that

80 $$X^{-1}Y = X_-^{-1}\, \text{diag}[s^{-r_i}]Y \in \mathbb{R}_p(s)^{n_i \times n_o}.$$

Hence by (79)-(80) we have shown that conditions (a)-(b) imply the existence of a solution s.t. (33) and (62) hold. ∎

Condition (66) of Theorem 61 shows that in order that the solution (X, Y) be int. proper, the row degrees of Y must be suitably bounded. The next theorem gives sufficient conditions for the existence of such solutions.

84 __Theorem__ [Existence of int. proper solutions of comp. eqn. (32)]. Consider the plant specifications (2)-(4) and the polynomial matrix $D_k \in \mathbb{R}[s]^{n_i \times n_i}$.

85 Let $(D_{p\ell}, N_{p\ell})$ be any int. pr. ℓ.c.f. of P with $D_{p\ell}$ row-reduced. Then the comp. eqn. (32) has a solution (X, Y) s.t.

33 $$(X, Y) \in \mathbb{R}[s]^{n_i \times n_i} \times \mathbb{R}[s]^{n_i \times n_o}$$

and

34 the ℓ.f. (X, Y) is int. pr. with X row-reduced with row degrees r_i, $i \in \underline{n}_i$,

if

63 $D_k \in \mathbb{R}[s]^{n_i \times n_i}$ is r.c.r. with row powers r_i, $i \in \underline{n}_i$, and column powers k_j, $j \in \underline{n}_i$,

s.t.

86 $$r_i \geq \mu - 1 \quad \forall i \in \underline{n}_i,$$

where

87 $$\mu := \max\{\partial_{ri}[D_{p\ell}],\ i \in \underline{n}_o\},$$

or equiv.

88 μ is the maximal degree of any entry of $D_{p\ell}$ where $D_{p\ell}$ is the left plant denominator in (85).

89 Comments. (a) Note that an int. pr. ℓ.c.f. $(D_{p\ell}, N_{p\ell})$ of P with $D_{p\ell}$ row-reduced is available in (36)-(38). Note also by Comment 2.4.3.33.L, that the row degrees of $D_{p\ell}$ are unique for a given plant P. In fact μ, as given by (88), is the observability index of the plant P [Kai. 1, Sec. 6.4.3, p. 413; Sec. 6.4.6, p. 431]; i.e., let [A, B, C, 0] be any minimal realization of P, then μ is the smallest integer ℓ s.t.

$$\mathrm{rk}\begin{bmatrix} C \\ CA \\ \vdots \\ CA^{\ell-1} \end{bmatrix} = \partial[\det(sI-A)] = \partial[\det D_{p\ell}].$$

or equiv. μ-1 is the minimum number of derivatives of the output needed to reconstruct all states at t = 0.

(b) Theorem 84 tells us that for solving problem COMP a safe choice of D_k is such that (63) and (86) hold. This implies that $\partial[\chi] = \partial[\det D_k] \geq \sum_{i=1}^{n_i} \{k_i + (\mu-1)\} = \partial[\det D_{pr}] + n_i(\mu-1)$. Compare with [Kai.1, Sec. 7.5, p. 535]. This implies also that the order of the compensator satisfies $\partial[\det D_{c\ell}] = \partial[\det X] \geq n_i(\mu-1)$.

(c) Condition (86) may be very conservative (see [Emr.1]): smaller row powers r_i may be acceptable for satisfying the conditions of Theorem 61 characterizing "internally proper" solutions.

91 Proof of Theorem 84. Return to Theorem 39 characterizing all polynomial matrix solutions (X, Y) of eqn. (32) . These solutions, for a D_k satisfying

(63) and (86), are given by

41 $$X = D_k V_{pr} - N_k N_{p\ell},$$

42 $$Y = D_k U_{pr} + N_k D_{p\ell}.$$

Now, in (42), we may consider Y as the result of the division of $D_k U_{pr}$ on the right by $D_{p\ell}$, i.e.,

$$D_k U_{pr} = -N_k D_{p\ell} + Y \quad \text{s.t.} \quad Y D_{p\ell}^{-1} \in \mathbb{R}_{p,o}(s)^{n_i \times n_o}.$$

Hence, by Fact 2.4.3.4.R,

$$\partial_{ci}[Y] < \partial_{ci}[D_{p\ell}] \quad \forall i \in \underline{n}_o.$$

Hence $\forall i \in \underline{n}_i$, $j \in \underline{n}_o$

$$\partial_{ri}[Y] \le \max \partial_{ri}[Y]$$

$$= \max \partial_{cj}[Y] < \max \partial_{cj}[D_{p\ell}]$$

$$= \max \partial_{rj}[D_{p\ell}] =: \mu.$$

Therefore, using (86),

$$\partial_{ri}[Y] \le \mu - 1 \le r_i \quad \forall i \in \underline{n}_i.$$

Hence we satisfy condition (66) of Theorem 61. Now, since condition (63) holds by assumption, all conditions of Theorem 61 hold and the comp. eqn. (32) must have a solution (X, Y) s.t. (33)-(34) holds. ∎

93 Summary for design. From a designer's point of view the results above can be summarized as follows:

$$C := D_{c\ell}^{-1} N_{c\ell} \in \mathbb{R}_p(s)^{n_i \times n_o} \text{ solves problem COMP}$$

if and only if

$(D_{c\ell}, N_{c\ell}) := (X, Y)$ is a solution of the comp. eqn.

32 $$XD_{pr} + YN_{pr} = D_k,$$

where

(a)

94 $$\det D_k(s) = \chi(s) \quad \text{with} \quad Z[\chi] \subset \mathbb{C}_-,$$

where i) χ is the prescribed char. poly. of Σ, and ii) D_k has the dynamical interpretation of Fact 30;

(b)

63 D_k is r.c.r. with row powers r_i, $i \in \underline{n}_i$ and column powers k_j, $j \in \underline{n}_i$,

where (i) the row powers must be supplied, (ii) the column powers are given by the plant in (3), and (iii) the highest degree coefficient matrix D_{kh} specifies X_h by (67) and (4);

(c)

66 $$\partial_{ri}[Y] \leq r_i \quad \forall i \in \underline{n}_i.$$

95 <u>Comment</u>. The summary above suggests also the following parametrization of all int. pr. solutions (X, Y) of (32) for a D_k satisfying (94) and (63): $(X, Y) \in \mathbb{R}[s]^{n_i \times n_i} \times \mathbb{R}[s]^{n_i \times n_o}$ must be s.t. (a) Y is given by the conditions

96 $$D_{pr} \text{ divides } D_k - YN_{pr} \text{ on the right,}$$

66 $$\partial_{ri}[Y] \leq r_i \quad \forall i \in \underline{n}_i,$$

and (b) X is given by the condition

97 $$D_k - YN_{pr} = XD_{pr}.$$

Since X is determined as a quotient by (97) once Y is given, it follows finally that all internally proper solutions (X, Y) of (32) for the given D_k are parameterized by Y satisfying (96) and (66).

A further transformation of the conditions (a) and (b) is possible. Let

$$n_p := \sum_{j=1}^{n_i} k_j = \partial[\det D_{pr}],$$

$$n_c := \sum_{i=1}^{n_i} r_i,$$

and assume, for reasons of simplicity, that

$$D_{pr} \in \mathbb{R}[s]^{n_i \times n_i} \text{ has } n_p \text{ distinct eigenvalues } p_q,\ q \in \underline{n}_p.$$

Hence D_{pr} has n_p nonzero eigenvectors $\ell_q \in \mathbb{C}^{n_i}$, $q \in \underline{n}_p$ s.t.

$$D_{pr}(p_q)\ell_q = \theta \quad \forall q \in \underline{n}_p.$$

Then, by [Goh.2, Th. 4.2], (96) is equivalent to

$$D_k(p_q)\ell_q = Y(p_q)N_{pr}(p_q)\ell_q \quad \forall q \in \underline{n}_p,$$

and (66) is equivalent to

$$Y(s) = S(s)Y_c,$$

where

$$S(s) := \text{block diag}\left[[s^{r_i}, s^{r_i-1}, \cdots, 1],\ i \in \underline{n}_i\right] \in \mathbb{R}[s]^{n_i \times (n_c+n_i)},$$

$$Y_c \in \mathbb{R}^{(n_c+n_i)\times n_o}.$$

Note that matrix Y_c has as entries the coefficients of the polynomial entries of $Y(\cdot) \in \mathbb{R}[s]^{n_i \times n_o}$. Finally, let $A \otimes B := [a_{ij}\, B]_{i,j}$ denote the Kronecker product of two matrices and $\sigma[Y_c] \in \mathbb{R}^{n_o(n_c+n_i)}$ denote a vector obtained by stacking the columns of Y_c. Then one obtains finally that conditions (96) and (66) are equivalent to the following linear equation in $\sigma[Y_c]$:

98 $$[D_k(p_q)\ell_q]_{q=1}^{n_p} = [(N_{pr}(p_q)\ell_q)^T \otimes S(p_q)]_{q=1}^{n_p}\, \sigma[Y_c],$$

where the vector on the LHS is of dimension $n_i \cdot n_p$ and the matrix on the RHS has dimension $n_i n_p \times n_o(n_c+n_i)$. Equation (98) parametrizes all Y satisfying (96) and (66) and hence all int. pr. solutions (X, Y) of (32) for the given D_k.

Chapter 7. Asymptotic Tracking

7.1. Introduction

This chapter develops first the theory of asymptotic tracking of a given class of inputs by a unity feedback system, where the forward path consists of a compensator followed by a given plant. Next the theory is used to design a tracking compensator.

Section 2 on theory starts with the analysis (7.2.4) of the tracking error of an exponentially stable unity feedback system. After defining (7.2.16) the class Ψ of inputs to be tracked and describing (7.2.19) the plant assumptions, the problem of asymptotic tracking is defined and discussed (7.2.25 et seg.). This leads (Theorem 7.2.31) to necessary conditions for asymptotic tracking on the compensator-plant forward path: essentially the forward path transfer function must be of full normal rank and may not have any zeros at the input characteristic frequencies. Theorem 7.2.60 develops sufficient conditions for asymptotic tracking: they are (1) exponential stability of the unity feedback system and (2) the factor condition: every entry of the compensator denominator (generated by a right coprime fraction) must have the input characteristic polynomial as a factor. (The latter condition is also known as the "internal model principle.") In Remark 7.2.73 we show that the problem of asymptotic disturbance rejection is almost equivalent to that of asymptotic tracking. In Theorem 7.2.75 we discuss the robustness of asymptotic tracking and disturbance rejection under the conditions of Theorem 7.2.60. Finally, Corollary 7.2.80 reformulates the tracking conditions in a form useful for design.

Section 3 treats the design of a tracking compensator. The problem is described (7.3.1), and the solution is given by an algorithm whose main step consists of solving the tracking compensator equation (7.3.18). This equation has as solution the polynomial matrix numerator and denominator of the tracking compensator. Theorem 7.3.33 describes the class of all polynomial matrix solutions. Finally, Theorem 7.3.45 characterizes internally proper solutions, while Theorem 7.3.55 gives a sufficient condition for their existence.

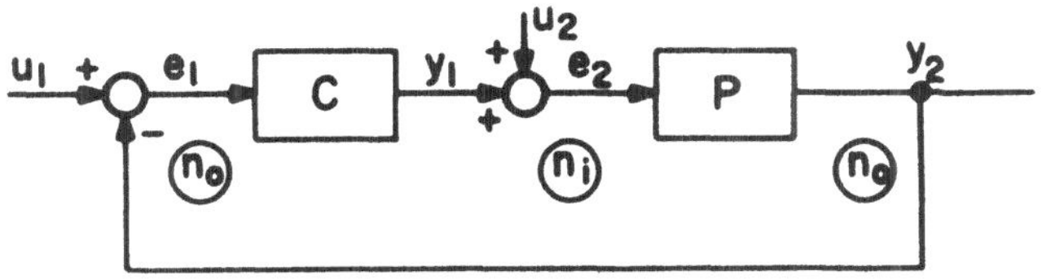

Fig. 4.3.1. The feedback system Σ under consideration.

7.2. Theory of Asymptotic Tracking

We study again feedback system Σ, shown in Fig. 4.3.1. As before P is the plant, C is the compensator, but now, for convenience, we call u_1, e_1, and y_2 resp. the input, error, and output of Σ.

1 Assumption. In feedback system Σ, shown in Fig. 4.3.1, the plant P and compensator C are such that

2 $P \in \mathbb{R}_{p,o}(s)^{n_o \times n_i}$ has an int. pr. ℓ.c.f. $(D_{p\ell}, N_{p\ell})$ and an int. pr. r.c.f. (N_{pr}, D_{pr}),

3 $C \in \mathbb{R}_p(s)^{n_i \times n_o}$ has an int. pr. ℓ.c.f. $(D_{c\ell}, N_{c\ell})$ and an int. pr. r.c.f. (N_{cr}, D_{cr}).

4 Analysis. Under Assumption 1, Theorem 4.3.6 on feedback system stability holds. In particular, with $P = D_{p\ell}^{-1}N_{p\ell}$, $C = N_{cr}D_{cr}^{-1}$, and $u_2 \equiv \theta$, the error e_1 of Σ is described by the well-formed PMD $\mathcal{D}_{e_1u_1} = [D_{p\ell}D_{cr} + N_{p\ell}N_{cr}, D_{p\ell}, D_{cr}, 0]$, i.e., by the equations

5 $$[D_{p\ell}D_{cr} + N_{p\ell}N_{cr}](p)\xi_c(t) = D_{p\ell}(p)u_1(t)$$
$$\forall t \geq 0,$$
6 $$e_1(t) = D_{cr}(p)\xi_c(t)$$

where $\xi_c(\cdot)$ is the compensator pseudo-state; moreover, $\xi_p(\cdot) = y_2(\cdot)$, the plant-pseudo-state, is then given by

7 $$\xi_p(t) = -D_{cr}(p)\xi_c(t) + u_1(t) \qquad \forall t \geq 0.$$

Hence, using the char. poly. of Σ, viz.,

8 $$\chi(s) := \det[D_{p\ell}D_{cr} + N_{p\ell}N_{cr}](s)$$

given by (4.3.9), and the input-error transfer function $H_{e_1u_1}$ given by

9 $$H_{e_1u_1} = D_{cr}[D_{p\ell}D_{cr} + N_{p\ell}N_{cr}]^{-1}D_{p\ell} \in \mathbb{R}_p(s)^{n_o \times n_o},$$

it follows that, for every initial value at $t = 0-$ of $\xi_c(\cdot)$ and $\xi_p(\cdot)$ and their derivatives, and for every p. suff. diff. $u_1(\cdot)$ with $u_1^{(j)}(0-) = \theta$ $\forall j = 0, 1, 2, \cdots$, $e_1(\cdot)$ shall be p. suff. diff. on $t \geq 0$ with Laplace transform

10 $$\hat{e}_1(s) = H_{e_1u_1}(s)\hat{u}_1(s) + \chi(s)^{-1}\, m(s),$$

where

11 $$m(s) \in \mathbb{R}[s]^{n_o}$$

is a polynomial vector depending on the initial values $\xi_c^{(j)}(0-)$ for $j = 0, 1, 2, \cdots$, s.t.

12 $$\chi(s)^{-1}m(s) \in \mathbb{R}_{p,o}(s)^{n_o}.$$

Moreover,

13 $$\hat{u}_1 \in \mathbb{R}_{p,o}(s)^{n_o} \Rightarrow \hat{e}_1 \in \mathbb{R}_{p,o}(s)^{n_o},$$ ∎

14 Exercise. Show that under Assumption 1, with $u_2 \equiv \theta$, the output y_2 of Σ is described by the PMD $\mathcal{D}_{y_2u_1} := [(D_{c\ell}D_{pr} + N_{c\ell}N_{pr}), N_{c\ell}, N_{pr}, 0]$.

In this section we propose to study the problem of asymptotic tracking under the following additional assumptions

16 Assumption. The class of inputs u_1 of Σ to be tracked is specified in terms of the Laplace transform as follows:

17 $$\Psi := \{\hat{u}_1 = \psi^{-1}\mu : \psi \in \mathbb{R}[s],\ \mu \in \mathbb{R}[s]^{n_o},\ \psi^{-1}\,\mu \in \mathbb{R}_{p,o}(s)^{n_o}\},$$

where ψ is a given monic polynomial and μ is arbitrary; furthermore,

18 $$Z[\psi] \subset \mathbb{C}_+.$$

19 Assumption. The plant $P \in \mathbb{R}_{p,o}(s)^{n_o \times n_i}$ is such that

20 $$n_o \leq n_i$$

and

21 $$Z[\psi] \cap Z[P] = \phi.$$ ∎

The problem of asymptotic tracking is now described by

25 Definition. We say that feedback system Σ, shown in Fig. 4.3.1, tracks asymptotically every input u_1 of class Ψ iff, with $u_2 \equiv \theta$, for every initial value at $t = 0-$ of the plant and compensator pseudo-states $\xi_p(\cdot)$ and $\xi_c(\cdot)$ and their derivatives, and for every input $u_1 \in \Psi$, with $u_1(t) = \theta_{n_o}$ for $t < 0$, we have that

$$e_1(t) \to \theta_{n_o} \text{ as } t \to \infty$$

or equiv.

$$y_2(t) - u_1(t) \to \theta_{n_o} \text{ as } t \to \infty.$$ ∎

26 Comment. For brevity, call the inputs $u_1 \in \Psi$ the reference inputs. They are described by (17)-(18) and are characterized as follows: Let

$$\psi(s) := \prod_{\ell=1}^{m} (s - \zeta_\ell)^{k_\ell},$$

where ζ_ℓ, for $\ell = 1, \cdots, m$, denotes the distinct roots of $\psi(\cdot)$. Let $1(t)$ denote the unit step function.

U.t.c.

(a)

27 $$u_1 \in \Psi$$

iff

$$u_1(t) = 1(t) \sum_{\ell=1}^{m} \sum_{q=0}^{k_\ell - 1} u_{\ell q} t^q e^{\zeta_\ell t} \quad \forall t,$$

where $\forall \ell$, q, $u_{\ell q} \in \mathbb{C}^{n_o}$ is arbitrary except for the fact that $u_{\ell q} = \bar{u}_{pq}$ if $\zeta_\ell = \bar{\zeta}_p \in \mathbb{C}\backslash\mathbb{R}$ and $u_{\ell q} \in \mathbb{R}^{n_o}$ if $\zeta_\ell \in \mathbb{R}$.

(b)

28 $$u_1 \in \Psi$$

iff

$$u_1(t) = \theta_{n_o} \quad \text{for} \quad t < 0$$

and for $t \geq 0$ <u>every component</u> of $u_1(\cdot)$ is a solution of the scalar differential equation

$$\psi(p)x(t) = 0 \qquad t \geq 0$$

for arbitrary initial conditions on $x(\cdot)$ and its derivatives at $t = 0$. ∎

In other words, the class Ψ is a set of linear combinations of waveforms of fixed shape.

Well-known waveforms to be tracked are bounded waveforms such as steps and sinusoids, and unbounded waveforms such as ramps, parabolas and increasing exponentials, justifying the requirement $Z[\psi] \subset \mathbb{C}_+$. Note that a bounded waveform going to θ_{n_o} as $t \to \infty$ is always tracked once the feedback system Σ is exp. stable (see Th. 3.3.2.19).

29 <u>Exercise</u>. Consider an exp. stable feedback system Σ satisfying Assumptions 1, 16, and 19. Assume that Σ tracks asymptotically every input of the class Ψ. Let $v_1 : \mathbb{R}_+ \to \mathbb{R}^{n_o}$ be bounded on compact intervals. Let $v_1(t) = \theta_{n_o}$ for $t < 0$, and, for some $u_1 \in \Psi$, $v_1(t) - u_1(t) \to \theta_{n_o}$, as $t \to \infty$. Show that, for any initial conditions at $t = 0-$ on $\xi_c(\cdot)$ and $\xi_p(\cdot)$ and their derivatives, the output due to these initial conditions and to $v_1(\cdot)$ satisfies $y_2(t) - u_1(t) \to \theta_{n_o}$.

30 <u>Comment</u>. The problem of asymptotic tracking occurs in practice in the following contexts.

(a) <u>Regulators</u> (e.g., Manufacturing processes: control of position, velocity, temperature, pH, ···). The problem is to track a set point which may be

changed from time to time (hence step inputs).
(b) Servomechanisms. For example, a radar plotting board following an airplane: the horizontal position of an airplane is determined by its polar coordinates with respect to the radar: the angle $\alpha(\cdot)$ (azimuth) and the distance $d(\cdot)$. The requirement is to track $\alpha(\cdot)$ and $d(\cdot)$, two slowly varying functions of time. The idea is to use polynomial interpolation: e.g., $\alpha(t) \simeq \alpha_0 + \alpha_1 t + \alpha_2 t^2$, a parabola, or $\alpha(t) \simeq \alpha_0 + \alpha_1 t + \alpha_2 t^2 + \alpha_3 t^3$, a cubic, and to require that the feedback system tracks any such polynomial sufficiently fast with zero asymptotic error. Note that an arbitrary parabola is a solution of $\dddot{x} = 0$.

A first task is to show that assumptions (20) and (21) are not restrictive.

31 Theorem [Necessary conditions for asymptotic tracking]. Let Assumption 1 hold and let Ψ be the class of reference inputs u_1 described by (17)-(18). Let Σ be exp. stable.
U.t.c., if

Σ tracks asymptotically every input $u_1 \in \Psi$,

then

32 $$\forall \zeta \in Z[\psi] \quad \det[N_{p\ell} N_{cr}](\zeta) \neq 0,$$

whence

33 $$n_o \leq n_i,$$

and

34 $$\forall \zeta \in Z[\psi] \quad rk[N_{p\ell}](\zeta) = n_o,$$

and

35 $$\forall \zeta \in Z[\psi] \quad rk[N_{cr}](\zeta) = n_o,$$

or equivalently:

with $PC \in \mathbb{R}_{p,o}(s)^{n_o \times n_o}$ denoting the forward path transfer function,

36 $$Z[PC] \cap Z[\psi] = \phi,$$

whence

37 $$n_o \leq n_i,$$

38 $$Z[P] \cap Z[\psi] = \phi,$$

and

39 $$Z[C] \cap Z[\psi] = \phi.$$

40 Comments. Theorem 31 teaches us that an exp. stable feedback system Σ will asymptotically track every input $u_1 \in \Psi$ only if the forward path transfer function PC has full normal rank n_o and no zeros at the input characteristic frequencies $\zeta \in Z[\psi]$. This implies that $n_o \leq n_i$ and that the plant and compensator have no zeros at any $\zeta \in Z[\psi]$. Hence assumptions (20) and (21) are necessary.

42 Proof of Theorem 31. (a) We shall first show the necessity of condition (32). By Assumption 1 feedback system Σ satisfies the assumptions of Theorem 4.3.6 and is exp. stable. Hence, with Definition 4.2.66, Σ is a well-formed PMD $\mathcal{D}_g$ which is exp. stable with char. poly.

43 $$\chi(s) := \det[D_{p\ell}D_{cr} + N_{p\ell}N_{cr}](s)$$

s.t.

44 $$Z[\chi] \subset \mathring{\mathbb{C}}_-.$$

By assumption also Σ tracks asymptotically every input $u_1 \in \Psi$ and, in particular, every input of the form

45 $$u_1(t) := u_o e^{\zeta t} \quad \text{where} \quad \zeta \in Z[\psi] \quad \text{and} \quad u_o \in \mathbb{C}^{n_o}.$$

Hence, with $u_2 \equiv \theta$, and by the exp. stability of Σ: $\forall \zeta \in Z[\psi]$, $\forall u_o \in \mathbb{C}^{n_o}$, $\forall t \geq 0$

46 $$y_2(t) = u_o e^{\zeta t} + \sum_i \sum_j y_{ij} t^j e^{\lambda_i t}$$

47 $$e_2(t) = y_1(t) = H_{y_1u_1}(\zeta) u_o e^{\zeta t} + \sum_i \sum_j e_{ij} t^j e^{\lambda_i t},$$

where (a) the λ_i's are (closed-loop) eigenvalues of Σ, whence $\mathrm{Re}\,\lambda_i < 0\ \forall i$ and (b) the y_{ij}'s and e_{ij}'s are constant vectors which depend on the initial conditions and the input.

Note further that

48 (i) $\qquad \forall \zeta \in Z[\psi] \quad H_{y_1u_1}(\zeta) = [N_{cr}(D_{p\ell}D_{cr} + N_{p\ell}N_{cr})^{-1}D_{p\ell}](\zeta) \in \mathbb{C}^{n_i \times n_o}$

because of (43)-(44) and (18), and

49 (ii) $\qquad D_{p\ell}(p)y_2(t) = N_{p\ell}(p)e_2(t) \quad \forall t \geq 0,$

where the latter equation is due to the plant PMD $\mathcal{D}_p = [D_{p\ell}, N_{p\ell}, I, 0]$ with $y_2(\cdot) = \xi_p(\cdot)$. Therefore, using (46)-(49) and letting $t \to \infty$, we obtain

$$\forall \zeta \in Z[\psi] \quad \forall u_o \in \mathbb{C}^{n_o}$$

50 $\qquad D_{p\ell}(\zeta)u_o = (N_{p\ell}N_{cr})(\zeta)\ \{(D_{p\ell}D_{cr} + N_{p\ell}N_{cr})^{-1}(\zeta)\}\ D_{p\ell}(\zeta)u_o,$

an equation which is well defined because of (43)-(44) and (18). Assume now for the purpose of a contradiction that (32) does not hold, i.e.,

51 $\qquad \exists\ \bar{\zeta} \in Z[\psi] \quad \text{s.t.} \quad \det[N_{p\ell}N_{cr}](\bar{\zeta}) = 0.$

or equiv.

52 $\qquad \exists\ \bar{\zeta} \in Z[\psi]$ and $\exists$ a nonzero $\gamma \in \mathbb{C}^{n_o}$ s.t. $\gamma^*(N_{p\ell}N_{cr})(\bar{\zeta}) = \theta^*.$

Hence in view of (50) and (52),

53 $\qquad \forall u_o \in \mathbb{C}^{n_o} \quad \gamma^* D_{p\ell}(\bar{\zeta})\ u_o = 0.$

Note now that, since Σ is exp. stable, by the analysis done in (4), the PMD $\mathcal{D}_{e_1u_1} = [D_{p\ell}D_{cr} + N_{p\ell}N_{cr}, D_{p\ell}, D_{cr}, 0]$ is exp. stable, and therefore does not have any <u>unstable</u> hidden modes. Hence, using some e.o.'s, we obtain

54 $\qquad \text{rk}[N_{p\ell}N_{cr} \,\vdots\, D_{p\ell}](s) = n_o \quad \forall s \in \mathbb{C}_+,$

55 $\qquad \text{rk}\begin{bmatrix} N_{p\ell}\ N_{cr} \\ \text{-------} \\ D_{cr} \end{bmatrix}(s) = n_o \quad \forall s \in \mathbb{C}_+,$

It follows now by (52), (54), and $\bar{\zeta} \in Z[\psi] \subset \mathbb{C}_+$ that

$$\eta^* := \gamma^* D_{p\ell}(\bar{\zeta}) \neq \theta^*.$$

However, setting $u_o = \eta$ in (53), we must also have

$$\eta^*\eta = 0, \text{ equiv. } \eta = \theta.$$

Hence we have reached a contradiction and assumption (51) is false, or equiv. (32) holds.

(b) We now establish the necessity of (33)-(35). We have that $N_{p\ell} \in \mathbb{R}[s]^{n_o \times n_i}$ and $N_{cr} \in \mathbb{R}[s]^{n_i \times n_o}$, whence by (32) and Sylvester's rule $\forall \zeta \in Z[\psi]$

$$n_o = rk[N_{p\ell}N_{cr}](\zeta) \leq \min\{rk[N_{p\ell}](\zeta),\ rk[N_{cr}](\zeta)\}$$

$$\leq \min(n_o, n_i) \leq n_i.$$

Hence (33)-(35) hold.

(c) The equivalence between (32) and (36) follows from the fact that by (54)-(55) $D_{p\ell}^{-1} N_{p\ell} N_{cr} D_{cr}^{-1}$ is a left-right fraction of PC with no unstable hidden modes, hence $Z[PC] \cap \mathbb{C}_+ = Z[N_{p\ell}N_{cr}] \cap \mathbb{C}_+$.

(d) Conditions (38) and (39) follow from (33)-(35) and Theorem 2.4.4.6. ∎

56 <u>Exercise</u>. Consider feedback system Σ (see Fig. 4.3.1), which satisfies Assumption 1. Let (N_r, D_r) be an int. pr. r.c.f. of PC; hence $H_{e_1u_1} = D_r(D_r + N_r)^{-1}$. Assume that Σ is exp. stable. Show that if Σ tracks asymptotically every input of class Ψ defined in (16), then

(a) $H_{e_1u_1}(s) = \psi(s)R(s)$ for some $R(s) \in \mathbb{R}_p(s)^{n_o \times n_o}$ s.t. $R(s)$ is analytic in $\mathbb{C}_+$;

(b) $D_r(s) = \psi(s)D(s)$ for some $D(s) \in \mathbb{R}[s]^{n_o \times n_o}$;

(c) $\det N_r(\zeta) \neq 0$, $\forall \zeta \in Z[\psi]$.

(Hint: Note Exercise 3.3.2.28.) ∎

We have now the main result.

60 <u>Theorem</u> [Sufficient condition for asymptotic tracking]. Consider feedback system Σ, shown in Fig. 4.3.1, under the system Assumptions 1, the input Assumptions 16, and the plant Assumptions 19.

U.t.c., <u>if</u>

(a)

61 feedback system Σ is exp. stable

or equiv.

62 $\chi(s) := \det[D_{p\ell}D_{cr} + N_{p\ell}N_{cr}](s)$ is s.t. $Z[\chi] \subset \mathring{\mathbb{C}}_-$;

(b)

63 $\psi \in \mathbb{R}[s]$ is a factor of every entry of D_{cr},

or equiv.

64 $D_{cr}(s) = \psi(s)D_c(s)$ for some $D_c \in \mathbb{R}[s]^{n_o \times n_o}$;

<u>then</u>

65 feedback system Σ tracks asymptotically every input $u_1 \in \Psi$.

66 <u>Comment</u>. Condition (63)⇔(64) is known as the "internal model principle" and states that the compensator $C = N_{cr}D_{cr}^{-1} = N_{cr}D_c^{-1}(\psi I_{n_o})^{-1}$ may be viewed as a series connection of two blocks, the first block "reproducing the input dynamics" in n_o parallel uncoupled channels. In particular, if we want to track steps, then one integrator must be present in every channel.

68 <u>Proof of Theorem 60</u>. Under Assumption 1, Theorem 4.3.6 on feedback system stability holds. Hence (62) is a necessary and sufficient condition for exp. stability. By the analysis done in (4), with $u_2 \equiv \theta$, for every initial value at $t = 0-$ of the plant and compensator pseudo-states $\xi_p(\cdot)$ resp. $\xi_c(\cdot)$ and their derivatives and for every p. suff. diff. $u_1(\cdot)$ with $u_1^{(j)}(0-) = \theta$ $\forall j = 0, 1, 2, \cdots$, the error $e_1(\cdot)$ shall be p. suff. diff. on $t \geq 0$ with Laplace transform

10 $$\hat{e}_1(s) = H_{e_1u_1}(s)\hat{u}_1(s) + \chi(s)^{-1}m(s),$$

where

11 $$m(s) \in \mathbb{R}[s]^{n_o}$$

is a polynomial vector depending on the initial values $\xi_c^{(j)}(0-)$ for

$j = 0, 1, 2, \cdots$ s.t.

12 $$\chi(s)^{-1} m(s) \in \mathbb{R}_{p,o}(s)^{n_o}.$$

Moreover,

9 $$H_{e_1u_1} = D_{cr}[D_{p\ell}D_{cr} + N_{p\ell}N_{cr}]^{-1}D_{p\ell} \in \mathbb{R}_p(s)^{n_o \times n_o}$$

and

13 $$\hat{u}_1 \in \mathbb{R}_{p,o}(s)^{n_o} \Rightarrow \hat{e}_1 \in \mathbb{R}_{p,o}(s)^{n_o}.$$

Now using (63)⇔(64), we obtain

69 $$H_{e_1u_1} = D_c[D_{p\ell}D_{cr} + N_{p\ell}N_{cr}]^{-1}D_{p\ell}\psi \in \mathbb{R}_p(s)^{n_o \times n_o}$$

since the <u>scalar</u> polynomial ψ commutes with any matrix. Pick any $u_1 \in \Psi$, whence by (17)

70 $$\hat{u}_1(s) = \psi(s)^{-1}\mu(s) \in \mathbb{R}_{p,o}(s)^{n_o}.$$

Hence the substitution of (69)-(70) into (10), and (12)-(13), (61)⇔(62) result in

$$\hat{e}_1(s) = D_c(s)[D_{p\ell}D_{cr} + N_{p\ell}N_{cr}](s)^{-1}D_{p\ell}(s)\mu(s) + \chi(s)^{-1}m(s)$$

s.t.

$$\hat{e}_1 \in \mathbb{R}_{p,o}(s)^{n_o} \text{ and } P[\hat{e}_1] \subset \mathring{\mathbb{C}}_-.$$

As a consequence $e_1(t) \to 0$ as $t \to \infty$, and feedback Σ tracks asymptotically every input u_1 of class Ψ. ∎

71 <u>Exercise</u>. Let Assumptions 1, 16, and 19 hold. Let $n_i = n_o =: n$. Show that if $Z[\det(D_{c\ell}D_{pr} + N_{c\ell}N_{pr})(s)] \subset \mathring{\mathbb{C}}_-$ and if $D_{c\ell}D_{pr} = \psi\, D_r$ for some $D_r \in \mathbb{R}[s]^{n \times n}$, then Σ tracks asymptotically every input $\in \Psi$.
(Hint: (i) $H_{y_2u_1} = [I + \psi N_{c\ell}^{-1}D_rN_{pr}^{-1}]$ is valid in some small neighborhood N of $Z[\psi]$. (ii) Write $H_{y_2u_1}$ from (i) as $H_{y_2u_1} = [I + \psi R(s)]^{-1}$ and for N sufficiently small $H_{y_2u_1}(s) = I - \psi(s)R(s) + \psi(s)^2R(s)^2 - \cdots$. (iii)$\cdots$.)

73 <u>Remark: additive plant-output disturbance rejection</u>. Assume that in feedback system Σ, shown in Fig. 4.3.1, plant-output disturbances d_o occur as shown in Fig. 1. Assume also that these disturbances are of class Ψ described by (17)-(18). Note that now

74 $$H_{y_2 d_o} = H_{e_1 u_1} = (I + PC)^{-1}.$$

It follows then, by an analysis similar to the proof of Theorem 60, that, under the assumptions and conditions of Theorem 60, with $u_1 \equiv \theta$ and $u_2 \equiv \theta$. for every initial value at t = 0- of the plant and compensator pseudo-states $\xi_p(\cdot)$ resp. $\xi_c(\cdot)$ and their derivatives, and for every disturbance $d_o \in \Psi$ with $d_o(t) = \theta$ for $t < 0$, $y_2(t) \to \theta$ as $t \to \infty$. Hence under the assumptions and conditions of Theorem 60, feedback Σ will <u>asymptotically track</u> every input $u_1 \in \Psi$ and <u>asymptotically reject</u> every plant-output disturbance $d_o \in \Psi$.

75 <u>Theorem</u> [Robustness of asymptotic tracking and disturbance rejection]. Assume that the feedback system Σ satisfies all the assumptions stated in Theorem 60. Consider <u>arbitrary changes</u> in the plant P, more precisely $N_{p\ell} \leftarrow \tilde{N}_{p\ell}$, $D_{p\ell} \leftarrow \tilde{D}_{p\ell}$ s.t. $(\tilde{D}_{p\ell}, \tilde{N}_{p\ell})$ is an int. pr. ℓ.c.f., and <u>arbitrary changes</u> in the compensator C, more precisely, $N_{cr} \leftarrow \tilde{N}_{cr}$, $D_c \leftarrow \tilde{D}_c$ s.t. $(\tilde{N}_{cr}, \psi\tilde{D}_c)$ is an int. pr. r.c.f. Assume further that the exp. stability of Σ is preserved. U.t.c. the perturbed system $\tilde{\Sigma}$ has the asymptotic tracking properties (stated in Theorem 60) and the disturbance rejection properties (stated in Remark 73). ∎

<u>Proof</u>. Follows the same steps as the proof of Theorem 60. ∎

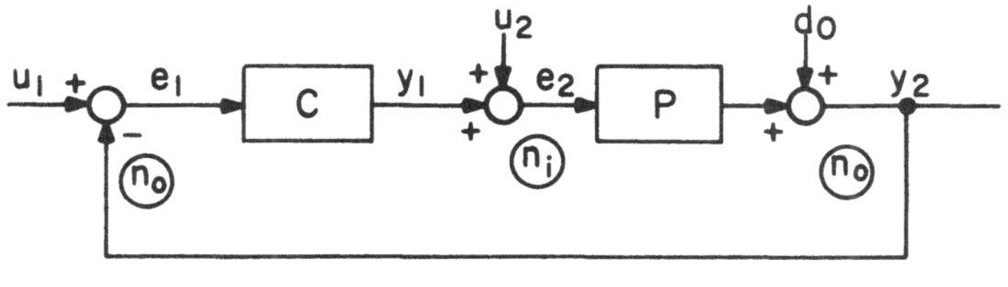

Fig. 1. Feedback system Σ with plant-output disturbances.

76 Comments. (a) The presence of the factor ψ in every element of D_{cr} (see (64)) is the crucial element in robust asymptotic tracking and disturbance rejection: note that P, D_c, and N_{cr} may undergo large perturbations (which must of course preserve exp. stability of Σ).
(b) The effectiveness of this factor ψ is made obvious if we refer to PMD $\mathcal{D}_{e_1 u_1}$ and eqn. (5) and (6). For good intuitive reasons, the zeros of ψ have been called blocking zeros of $H_{e_1 u_1}$ (see the scalar factor ψ in eqn. (69)).

80 Corollary. [Tracking for compensator design]. Consider feedback system Σ, shown in Fig. 4.3.1, where

81 (1) $n_o = n_i =: n$,

82 (2) $P \in \mathbb{R}_{p,o}(s)^{n\times n}$ has an int. pr. r.c.f. (N_{pr}, D_{pr}),

83 (3) $C \in \mathbb{R}_p(s)^{n\times n}$ has an int. pr. ℓ.f. $(D_{c\ell}, N_{c\ell})$,

84 (4) the input-assumption 16 and plant-assumptions 19 hold.
U.t.c., if

85 (a) feedback system Σ is exp. stable,

or equiv.

86 $$\chi(s) := \det[D_{c\ell}D_{pr} + N_{c\ell}N_{pr}](s) \text{ is s.t. } Z[\chi] \subset \mathring{\mathbb{C}}_-,$$

87 (b) ψ is a factor of every entry of $D_{c\ell}$,

or equiv.

88 $D_{c\ell}(s) = \psi(s)D_c(s)$ for some $D_c \in \mathbb{R}[s]^{n\times n}$,

then

89 feedback system Σ tracks asymptotically every input $u_1 \in \Psi$.

90 Comments (a) By choosing $P = N_{pr}\, D_{pr}^{-1}$ and $C = D_{c\ell}^{-1}\, N_{c\ell}$, one considers implicitly the I/O map

91 $$H_{y_2 u_1} = N_{pr}[D_{c\ell}D_{pr} + N_{c\ell}N_{pr}]^{-1}N_{c\ell}.$$

(b) Note also that the corollary does not require C to have a coprime fraction: this is useful because the proposed compensator design does not always deliver C as a coprime fraction. It will also follow from the analysis below that the factor condition (88) will be preserved after canceling common left factors.

(c) Since $Z[\psi] \subset \mathbb{C}_+$ it follows from the stability requirement (86) and the factor condition (88) that

92 $$\det N_{c\ell}(\zeta) \neq 0 \quad \text{and} \quad \det N_{pr}(\zeta) \neq 0 \qquad \forall \zeta \in Z[\psi].$$

We start the proof of Corollary 80 with three lemmas.

95 Lemma. Let D and $\bar{D}$ be two equivalent nonsingular polynomial matrices belonging to $\mathbb{R}[s]^{n\times n}$. Let $\psi \in \mathbb{R}[s]$ and let $\psi|D$ denote the fact that ψ is a factor of every entry of D.

U.t.c.

$$\psi|D \quad \Leftrightarrow \quad \psi|\bar{D}.$$

Proof. Since D and $\bar{D}$ are equivalent, $\exists$ unimodular matrices L and R in $\mathbb{R}[s]^{n\times n}$ s.t. $D = L\,\bar{D}\,R$ and $L^{-1}\,D\,R^{-1} = \bar{D}$, where L, L^{-1}, R, R^{-1} are polynomial matrices. Hence, if $D = \psi\tilde{D}$ for some $\tilde{D} \in \mathbb{R}[s]^{n\times n}$, then $\bar{D} = L^{-1}\psi\tilde{D}\,R^{-1}$, whence $\psi|D \Rightarrow \psi|\bar{D}$. Similarly, if $\bar{D} = \psi\check{D}$ for some $\check{D} \in \mathbb{R}[s]^{n\times n}$, then $D = L\psi\check{D}R$, whence $\psi|\bar{D} \Rightarrow \psi|D$. ∎

96 Lemma. Let $(D_{c\ell}, N_{c\ell})$ be any ℓ.f. of $C \in \mathbb{R}(s)^{n\times n}$. Let $\psi \in \mathbb{R}[s]$ s.t.

97 $$\det N_{c\ell}(\zeta) \neq 0 \qquad \forall \zeta \in Z[\psi].$$

Let (N_{cr}, D_{cr}) be any r.c.f. of $C \in \mathbb{R}(s)^{n\times n}$.

U.t.c.

98 $$\psi|D_{c\ell} \quad \Rightarrow \quad \psi|D_{cr}.$$

Proof. Let $(\bar{D}_{c\ell}, \bar{N}_{c\ell})$ be any ℓ.c.f. of $C \in \mathbb{R}(s)^{n\times n}$. Now since (N_{cr}, D_{cr}) is a r.c.f. of C and by Theorem 2.4.2.41 D_{cr} and $\bar{D}_{c\ell}$ are equivalent, it follows by Lemma 85 that $\psi|D_{cr} \Leftrightarrow \psi|\bar{D}_{c\ell}$. Hence (98) holds if

99 $$\psi|D_{c\ell} \Rightarrow \psi|\bar{D}_{c\ell}.$$

We show now that (99) holds.

Now if $\psi | D_{c\ell}$, then $D_{c\ell} = \psi D_c$ for some $D_c \in \mathbb{R}[s]^{n\times n}$. Now let $\tilde{L} \in \mathbb{R}[s]^{n\times n}$ be a g.c.ℓ.d. of $(D_c, N_{c\ell})$, whence $\exists\ \tilde{D}_c$ and $\tilde{N}_{c\ell}$ in $\mathbb{R}[s]^{n\times n}$ s.t.

100 $$D_c = \tilde{L}\,\tilde{D}_c,\ N_{c\ell} = \tilde{L}\,\tilde{N}_{c\ell},$$

101 and $(\tilde{D}_c, \tilde{N}_{c\ell})$ is ℓ.c.

Note now that by (100) and (97) $\det \tilde{N}_{c\ell}(\zeta) \neq 0 \quad \forall \zeta \in Z[\psi]$, whence by (101), $(\psi\tilde{D}_c, \tilde{N}_{c\ell})$ is ℓ.c. Moreover, since, by (100), $D_{c\ell} = \psi D_c = \tilde{L}\psi\tilde{D}_c$ and $N_{c\ell} = \tilde{L}\tilde{N}_{c\ell}$, we have $C = D_{c\ell}^{-1} N_{c\ell} = (\psi\tilde{D}_c)^{-1}\tilde{N}_{c\ell}$. Hence C has a ℓ.c.f. $(\psi\tilde{D}_c, \tilde{N}_{c\ell})$. Now, since $(\bar{D}_{c\ell}, \bar{N}_{c\ell})$ is also a ℓ.c.f. of C, it follows by Theorem 2.4.1.17.L that there exists a unimodular matrix L s.t. $L\psi\tilde{D}_c = \bar{D}_{c\ell}$. Therefore, $\psi | \bar{D}_{c\ell}$ and (99) holds. ∎

103 <u>Lemma</u>. Consider feedback system Σ, shown in Fig. 4.3.1, where
(a) $P \in \mathbb{R}_{p,o}(s)^{n_o\times n_i}$ has an int. pr. ℓ.c.f. $(D_{p\ell}, N_{p\ell})$ and an int. pr. r.c.f. (N_{pr}, D_{pr}),
(c) $C \in \mathbb{R}_p(s)^{n_i\times n_o}$ has an int. pr. ℓ.f. $(D_{c\ell}, N_{c\ell})$ and an int. pr. r.c.f. (N_{cr}, D_{cr}).
U.t.c.

104 $\det[D_{p\ell}\, D_{cr} + N_{p\ell}\, N_{cr}]$ is a factor of $\det[D_{c\ell}D_{pr} + N_{c\ell}N_{pr}]$.

<u>Proof</u>. Let $L \in E(\mathbb{R}[s])$ be any g.c.ℓ.d. of $(D_{c\ell}, N_{c\ell})$, whence there exist $\bar{D}_{c\ell}$ and $\bar{N}_{c\ell} \in E(\mathbb{R}[s])$ s.t. $D_{c\ell} = L\,\bar{D}_{c\ell}$, $N_{c\ell} = L\,\bar{N}_{c\ell}$ with $(\bar{D}_{c\ell}, \bar{N}_{c\ell})$ an int. pr. ℓ.c.f. of C. Hence,

105 $$\det[D_{c\ell}D_{pr} + N_{c\ell}N_{pr}] = \det L \det[\bar{D}_{c\ell}D_{pr} + \bar{N}_{c\ell}N_{pr}].$$

Observe that, with $P = N_{pr}\, D_{pr}^{-1} = D_{p\ell}^{-1}\, N_{p\ell}$ and $C = N_{cr}\, D_{cr}^{-1} = \bar{D}_{c\ell}^{-1}\, \bar{N}_{c\ell}$, the assumptions of Theorem 4.3.6 are satisfied, whence

106 $$\det[\bar{D}_{c\ell}D_{pr} + \bar{N}_{c\ell}N_{pr}] \sim \det[D_{p\ell}D_{cr} + N_{p\ell}N_{cr}].$$

Hence, in view of (105) and (106), (104) follows. ∎

110 Proof of Corollary 80. Notice that by assumptions (81)-(83) we satisfy the assumptions of feedback-stability Corollary 4.3.26. Hence (86) is a necessary and sufficient condition for the exp. stability of Σ. Therefore, under (81)-(83) and (86), the feedback system Σ is exp. stable. Moreover, with $u_2 \equiv \theta$, the error $e_1(\cdot)$ of Σ is described by the well-formed and exp. stable PMD $\mathcal{D}_{e_1u_1} = [D_{c\ell}D_{pr} + N_{c\ell}N_{pr}, N_{c\ell}, -N_{pr}, I]$ by the equations

$$\begin{aligned} [D_{c\ell}D_{pr} + N_{c\ell}N_{pr}](p)\xi_p(t) &= N_{c\ell}(p)u_1(t) \\ e_1(t) &= -N_{pr}(p)\xi_p(t) + u_1(t) \end{aligned} \qquad \forall t \geq 0,$$

where $\xi_p(\cdot)$ is the plant pseudo-state. Moreover, $y_1(\cdot) = \xi_c(\cdot)$, viz., the compensator-pseudo-state is given by

$$\xi_c(t) = D_{pr}(p)\xi_p(t) \qquad \forall t \geq 0.$$

Finally, the input-error transfer function $H_{e_1u_1}$ is given by

$$H_{e_1u_1} = I - H_{y_2u_1} = I - N_{pr}\,[D_{c\ell}D_{pr} + N_{c\ell}N_{pr}]^{-1}\,N_{c\ell} \in \mathbb{R}_p(s)^{n\times n}.$$

Let now $\hat{u}_1 = \psi^{-1}\mu \in \Psi$, where Ψ is given by (17)-(18); then, for every initial value at $t = 0-$ of $\xi_p(\cdot)$ and $\xi_c(\cdot)$ and their derivatives, the Laplace transform of the error $e_1(\cdot)$ of Σ is given by

111 $$\hat{e}_1 = H_{e_1u_1}\psi^{-1}\mu + \chi^{-1}m \in \mathbb{R}_{p,o}(s)^n,$$

where χ is the char. poly. (86) and $m \in \mathbb{R}[s]^n$ is a polynomial vector depending on the initial values at $t = 0-$ of $\xi_p(\cdot)$ and its derivatives.

Now let $(D_{p\ell}, N_{p\ell})$ be an int. pr. ℓ.c.f. of P and (N_{cr}, D_{cr}) an intr. r.c.f of C, then

112 $$H_{e_1u_1} = D_{cr}\,[D_{p\ell}D_{cr} + N_{p\ell}N_{cr}]^{-1}\,D_{p\ell} \in \mathbb{R}_p(s)^{n\times n},$$

where we note that because of Lemma 103,

113 $\det[D_{p\ell}D_{cr} + N_{p\ell}N_{cr}]$ is a factor of $\chi(s)$ in (86).

Moreover, since by stability condition (86) and factor condition (88) det $N_{c\ell}(\zeta) \neq 0 \quad \forall \zeta \in Z[\psi] \subset \mathbb{C}_+$, we have by Lemma 96 that

114 $$D_{cr} = \psi \tilde{D}_c \text{ for some } \tilde{D}_c \in \mathbb{R}[s]^{n\times n}.$$

Hence in view of (86), (111)-(114) we have

$$\hat{e}_1 = \tilde{D}_c[D_{p\ell}D_{cr} + N_{p\ell}N_{cr}]^{-1}D_{p\ell}\,\mu + \chi^{-1}m \in \mathbb{R}_{p,o}(s)^n,$$

where $P[\hat{e}_1] \subset \mathring{\mathbb{C}}_-$. Therefore, $\lim_{t\to\infty} e_1(t) = \theta$ and feedback system Σ tracks asymptotically every input u_1 of class Ψ. ∎

7.3. The Tracking Compensator Problem

1 Problem TRC. We are given

(a) a class of reference inputs u_1 to be tracked, specified in terms of the Laplace transform by

2 $$\Psi := \{\hat{u}_1 = \psi^{-1}\mu : \psi \in \mathbb{R}[s],\ \mu \in \mathbb{R}[s]^n,\ \psi^{-1}\mu \in \mathbb{R}_{p,o}(s)^n\},$$

where ψ is a given monic polynomial and μ is arbitrary; furthermore,

3 $$Z[\psi] \subset \mathbb{C}_+.$$

(b) a square (i.e., $n_o = n_i =: n$) plant P s.t.

4 $P \in \mathbb{R}_{p,o}(s)^{n\times n}$ has an int. pr. r.c.f. (N_{pr}, D_{pr}) s.t.

5 D_{pr} is column reduced with column degrees k_j, $j \in \underline{n}$, and

6 with highest column degree coefficient matrix $D_{ph} = I_n$;

7 $Z[P] \cap Z[\psi] = \phi$ or equiv. $\det[N_{pr}](\zeta) \neq 0 \quad \forall \zeta \in Z[\psi]$.

We must find a compensator C s.t.

8 $C \in \mathbb{R}_p(s)^{n\times n}$ has an int. pr. ℓ.f. $(D_{c\ell}, N_{c\ell})$ with

9 $D_{c\ell}$ row reduced,

s.t.

10 the feedback system Σ, shown in Fig. 4.3.1, is exp. stable with prescribed closed-loop characteristic polynomial $\chi(s)$ and

11 the feedback system Σ tracks asymptotically every input u_1 of class Ψ.

■

In view of Corollary 7.2.80, solutions of problem TRC are obtained by the following

15 Algorithm [Solution of problem TRC]

Step 1. Pick $\chi(s) \in \mathbb{R}[s]$ s.t.

16 $$Z[\chi] \subset \mathring{\mathbb{C}}_-.$$

Step 2. Pick $D_k \in \mathbb{R}[s]^{n\times n}$ s.t.

17 (a) $\det D_k(s) = \chi(s)$;

(b) the tracking compensator equation

18 $$X\psi D_{pr} + YN_{pr} = D_k$$

has a solution (X, Y) s.t.

19 $(X, Y) \in \mathbb{R}[s]^{n\times n} \times \mathbb{R}[s]^{n\times n}$

and

20 $(\psi X, Y)$ is an int. pr. ℓ.f. with ψX row reduced.

Step 3. Pick a solution (X, Y) of the tracking comp. eqn. (18) s.t. (19)-(20) holds.

Set

21 $$D_{c\ell} := \psi X \quad N_{c\ell} := Y$$

22 $$C := D_{c\ell}^{-1} N_{c\ell}.$$

■

23 Comments. (a) C given by (22) is a solution of problem TRC. Indeed, since $X(s) = \det[D_{c\ell}D_{pr} + N_{c\ell}N_{pr}](s)$, $Z[x] \subset \mathring{\mathbb{C}}_-$, and $D_{c\ell} = \psi X =: \psi D_c$ for some $D_c \in \mathbb{R}[s]^{n\times n}$, all conditions of Corollary 7.2.80 are met.
(b) Matrix D_k has the dynamical interpretation of Fact 6.2.30; the tracking comp. eqn. (18) is similar to comp. eqn. (6.2.32) since it will turn out that $(N_{pr}, \psi D_{pr})$ is an int. pr. r.c.f. of $\psi^{-1}P \in \mathbb{R}_{p,o}(s)^{n\times n}$.

Solving eqn. (18) for the tracking compensator. In design the input class Ψ (2)-(3) and the plant data (4)-(7) are given.

26 Fact. Given the data (2)-(7), we have that $(N_{pr}, \psi D_{pr})$ is an int. pr. r.c.f. of $\psi^{-1}P \in \mathbb{R}_{p,o}(s)^{n\times n}$.

Proof. By (4)-(5), $P \in \mathbb{R}_{p,o}(s)^{n\times n}$ has an int. pr. r.c.f. (N_{pr}, D_{pr}), with D_{pr} column reduced, and by (7) $\det N_{pr}(\zeta) \neq 0 \ \forall \zeta \in Z[\psi]$. Therefore,

$$\mathrm{rk}\begin{bmatrix} \psi D_{pr} \\ \hline N_{pr} \end{bmatrix}(s) = \mathrm{rk}\left[\begin{array}{c|c} \psi I & 0 \\ \hline 0 & I \end{array}\right]\begin{bmatrix} D_{pr} \\ \hline N_{pr} \end{bmatrix}(s) = n \quad \forall s \in \mathbb{C}\backslash Z[\psi]$$

and $\forall \zeta \in Z[\psi]$. Hence $(N_{pr}, \psi D_{pr})$ is a r.c.f. of $\psi^{-1}P \in \mathbb{R}_{p,o}(s)^{n\times n}$. Moreover, this fraction is int. proper since ψD_{pr} is column reduced. ∎

Note now that since, by Fact 26, $(N_{pr}, \psi D_{pr})$ is a r.c.f. of $\psi^{-1}P$, Theorem 2.4.1.25.R delivers a generalized Bezout identity: there exist six polynomial matrices

27 $$U_r,\ V_r,\ \psi D_{p\ell},\ N_{p\ell},\ U_\ell,\ V_\ell \in E(\mathbb{R}[s])$$

s.t.

28 $$\left[\begin{array}{c|c} V_r & U_r \\ \hline -N_{p\ell} & \psi D_{p\ell} \end{array}\right]\left[\begin{array}{c|c} \psi D_{pr} & -U_\ell \\ \hline N_{pr} & V_\ell \end{array}\right] = \left[\begin{array}{c|c} I_n & 0 \\ \hline 0 & I_n \end{array}\right].$$

Moreover,

29 $(\psi D_{p\ell}, N_{p\ell})$ is a ℓ.c.f. of $\psi^{-1}P$,

30 $(D_{p\ell}, N_{p\ell})$ is a ℓ.c.f. of P,

and without loss of generality

31 $\psi D_{p\ell}$ and $D_{p\ell}$ are row reduced.

Proof. If we first replace $\psi D_{p\ell}$ by $D_\ell \in \mathbb{R}[s]^{n\times n}$, then the existence of the resulting six matrices, listed in (27), s.t. (28)-(29) hold, is due to Theorem 2.4.1.25.R. Note in particular that one obtains $(N_{pr}, \psi D_{pr})$ is a r.c.f. and $(D_\ell, N_{p\ell})$ is a ℓ.c.f. of $\psi^{-1}P$, where the denominators ψD_{pr} and D_ℓ are equivalent polynomial matrices (observe that $\psi^{-1}P \in \mathbb{R}(s)^{n\times n}$ and use Theorem 2.4.2.41). Hence, by Lemma 7.2.95, $D_\ell = \psi D_{p\ell}$ for some $D_{p\ell} \in \mathbb{R}[s]^{n\times n}$. This observation justifies the existence of the six matrices (27) s.t. (28)-(31) hold. In particular, (30) holds because $\det N_{p\ell}(\zeta) \neq 0 \ \forall \zeta \in Z[\psi]$ (note that the first matrix on the LHS of (28) is unimodular). ∎

33 Theorem [Polynomial solutions of the tracking comp. eqn. (18)]. Consider eqn. (18), where (a) $P = N_{pr} D_{pr}^{-1}$ satisfies (4)-(7), (b) $\psi \in \mathbb{R}[s]$ is a monic polynomial s.t. (3) holds, and (c) $D_k \in \mathbb{R}[s]^{n\times n}$ is s.t. (16)-(17) holds. U.t.c.

34 $(X, Y) \in \mathbb{R}[s]^{n\times n} \times \mathbb{R}[s]^{n\times n}$ is a solution of (18)

$\Leftrightarrow$

$\exists\ N_k \in \mathbb{R}[s]^{n\times n}$ s.t.

35 $X = D_k V_r - N_k N_{p\ell}$

36 $Y = D_k U_r + N_k \psi D_{p\ell}$,

where U_r, V_r, $\psi D_{p\ell}$, $N_{p\ell}$ are elements of the generalized Bezout identity (28). Moreover,

37 $\det Y(\zeta) \neq 0 \quad \forall \zeta \in Z[\psi]$,

whence

38 $(\psi X, Y)$ is ℓ.c. $\Leftrightarrow$ (X, Y) is ℓ.c. $\Leftrightarrow$ (D_k, N_k) is ℓ.c. ∎

39 Comments. (a) Except for (37)-(38), the result is similar to that of Theorem 6.2.39 on the polynomial solutions of the comp. eqn. (6.2.32).
(b) Given a solution (X, Y), we set $(\psi X, Y) =: (D_{c\ell}, N_{c\ell})$ i.e., we obtain the compensator $C = D_{c\ell}^{-1}N_{c\ell} = (\psi X)^{-1}Y$ provided that $\det X \neq 0$. The additional

requirement that $(\psi X, Y)$ be an int. pr. ℓ.f. with ψX row-reduced is handled below.
(c) Condition (37) on $N_{c\ell} := Y$ is essential for tracking (see Comment 7.2.90(c)).

42 Proof of Theorem 33. The proof of (a) the equivalence between (34) and (35)-(36) and (b) (X, Y) is ℓ.c. ⇔ (D_k, N_k) is ℓ.c. uses the generalized Bezout identity (28) and repeats the steps of the proof of Theorem 6.2.39. Condition (37) holds because (a) in (28) $\det U_r(\zeta) \neq 0 \quad \forall \zeta \in Z[\psi]$ (exercise), (b) $\det D_k(\zeta) \neq 0 \quad \forall \zeta \in Z[\psi] \subset \mathbb{C}_+$, and (c) by (36)

$$\det Y(\zeta) = \det D_k(\zeta) \det U_r(\zeta) \neq 0 \quad \forall \zeta \in Z[\psi].$$

Finally by condition (37), $(\psi X, Y)$ is ℓ.c. ⇔ (X, Y) is ℓ.c. Hence we are done. ∎

We now give conditions under which the solutions (X, Y) of the tracking comp. eqn. (18) are s.t. $(\psi X)^{-1}Y$ is an int. proper left fraction. The proof of the result below is similar to the proof of Theorem 6.2.61 and is therefore omitted.

45 Theorem [Internally proper solutions of tracking comp. eqn. (18)]. Consider the tracking comp. eqn. (18), where (a) $P = N_{pr} D_{pr}^{-1}$ satisfies (4)-(7), (b) $\psi \in \mathbb{R}[s]$ is a monic polynomial s.t. (3) holds, and (c) $D_k \in \mathbb{R}[s]^{n \times n}$ s.t. (16)-(17) holds.
U.t.c.
The tracking compensator equation, given by

18 $$X\psi D_{pr} + YN_{pr} = D_k,$$

has a solution (X, Y) s.t.

19 $$(X, Y) \in \mathbb{R}[s]^{n \times n} \times \mathbb{R}[s]^{n \times n}$$

and

20 the left fraction $(\psi X, Y)$ is int. proper with ψX row reduced with row degrees r_i, $i \in \underline{n}$, and highest degree coefficient matrix X_h

if and only if

46 (a) D_k is row-column-reduced with row powers r_i, $i \in \underline{n}$, resp. column powers k_j, $j \in \underline{n}$,

or equiv.,

47 $D_k(s) = \text{diag}[s^{r_i}]_{i=1}^{n}\, D_{k-}(s)\, \text{diag}[s^{k_j}]_{j=1}^{n}$, where $D_{k-} \in \mathbb{R}(s)^{n\times n}$ is biproper and the highest degree coefficient matrix D_{kh} of D_k satisfies

48 $$D_{kh} = D_{k-}(\infty);$$

49 (b) for the given D_k, (X, Y) is a polynomial matrix solution of tracking comp. eqn. (18) s.t. the row degrees of Y satisfy

$$\partial_{ri}[Y] \le r_i \quad \forall i \in \underline{n}.$$

Moreover, under conditions (a)-(b) we have that the highest degree coefficient matrices of D_{pr}, ψX, and D_k are related by

50 $$D_{kh} = X_h\, D_{ph}$$

s.t., if $D_{ph} = I$ and $D_{kh} = I$, then $X_h = I$. ∎

51 Comments. (a) In (20), note that ψX is row reduced iff X is row reduced; moreover, with $\psi \in \mathbb{R}[s]$ a monic polynomial the highest row degree coefficient matrices of ψX and X are identical.

(b) From Comment 39(b) and the expanded statement (20), the order of the solving compensator C is $\sum_{i=1}^{n} r_i$, where the r_i's are the row powers of the chosen matrix D_k (see (46)). We shall see below that if these row powers r_i are sufficiently large, then a solution (X, Y) satisfying (49) will exist. Hence again we must choose $\chi(s) = \det D_k(s)$, the char. poly. of Σ, to be of sufficiently large degree for problem TRC to have a solution. ∎

A reasoning similar to that of Theorem 6.2.84 now leads to:

55 Theorem [Existence of int. pr. solutions of tracking comp. eqn. (18)]. Consider the tracking comp. eqn. (18) under the assumptions of Theorem 45.

56 Let $(D_{p\ell}, N_{p\ell})$ be any int. pr. ℓ.c.f. of P with $D_{p\ell}$ row reduced. Then the tracking comp. eqn. (18) has a solution (X, Y) s.t. (19)-(20) hold

if

46 $D_k \in \mathbb{R}[s]^{n\times n}$ is r.c.r. with row powers r_i, $i \in \underline{n}$, and column powers k_j, $j \in \underline{n}$,

s.t.

57 $$r_i \geq \partial\psi + \mu - 1 \quad \forall i \in \underline{n},$$

where

58 $$\mu := \max\{\partial_{ri}[D_{p\ell}],\ i \in \underline{n}\},$$

or equiv.,

59 μ is the maximal degree of any entry of $D_{p\ell}$ where $D_{p\ell}$ is the left plant denominator in (56). ∎

Proof. Observe that in (36) Y may be viewed as the remainder of the division of $D_k U_r$ on the right by $\psi D_{p\ell}$ and repeat the reasoning of the proof of Theorem 6.2.84. ∎

Chapter 8. Design with Stable Plants

8.1. Introduction

In this chapter, Sect. 2 develops the theory of unity feedback systems with stable plants. Section 3 uses this theory to achieve decoupling and eigenvalue placement. Section 4 shows how in the case of an unstable plant one may first stabilize it by local feedback.

In Sect. 2, a simple frequency domain analysis (8.2.6) displays a parameter matrix Q, in terms of which all closed-loop transfer functions are easily calculated. Furthermore, for a given exponentially stable plant, all proper and exp. stable Q's globally parametrize all proper compensators which result in an exp. stable unity feedback system (Theorem 8.2.30). Thus the main design task of providing an exp. stable feedback system and a proper compensator is automatically satisfied. We are left with the design task of meeting other specifications: e.g., achieving a small sensitivity over a given band without saturating the plant. There may be some other limitations on the achievable I/O maps: e.g., Theorem 8.2.45 proves that any plant-$\mathbb{C}_+$-zero is a $\mathbb{C}_+$-zero of the I/O map. Furthermore, Theorem 8.2.70 displays a lower bound on desensitization due to plant-$\mathbb{C}_+$-zeros.

Section 3 contains, for a given exp. stable plant, a Q-parameter algorithm for the design of a compensator, such that the resulting unity feedback system achieves a decoupled (equiv. diagonal) I/O map. This includes the assignment of closed-loop eigenvalues channel by channel.

Section 4 shows how the advantages of Q-parametrization can be extended to the case of unstable plants. Here an unstable plant is first locally stabilized, and thereafter a global parametrization of all exp. stable two-loop feedback systems is found similar to Theorem 8.2.30 (Th. 8.4.18). We obtain finally that all I/O maps, achievable by a single-loop exp. stable feedback system, can be achieved by the two-step scheme described above (Exercise 8.4.24).

8.2. Q-Parametrization Design Properties [Zam. 1, Des. 5]

In this chapter we shall consider the feedback system Σ, shown in Fig. 1, where P is the plant, C is the compensator, and where u_1, u_2, e_1, e_2, y_1, y_2

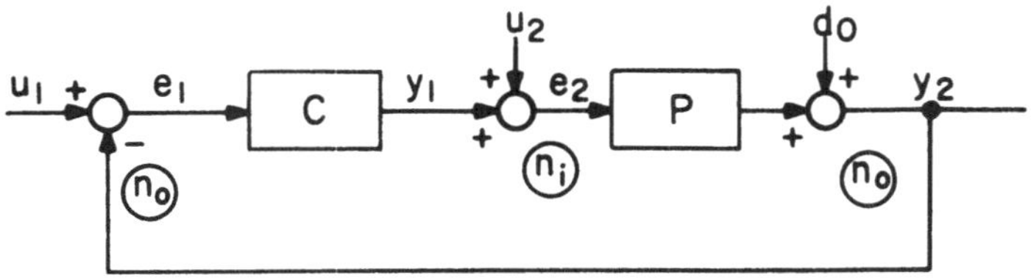

Fig. 1. The feedback system Σ under consideration.

are the usual inputs and outputs; moreover, d_o is a possible disturbance at the output of the plant

1 Assumptions. For feedback system Σ, it is assumed that

2 $P \in \mathbb{R}_{p,o}(s)^{n_o \times n_i}$

3^a $C \in \mathbb{R}(s)^{n_i \times n_o}$; furthermore, using P and C we can define the parameter $Q \in \mathbb{R}(s)^{n_i \times n_o}$ by

3^b $Q := C(I + PC)^{-1} = (I + CP)^{-1}C$;

we assume that

3^c $Q \in \mathbb{R}_p(s)^{n_i \times n_o}$. ∎

In Theorem 14, we shall show that (3^c) will guarantee that all closed-loop transfer functions of Σ are proper.

6 Analysis. As soon as (2) and (3^a) hold and $(I + PC)^{-1} \in E(\mathbb{R}(s))$, we can perform the following algebra:

7 $$I - PQ = (I + PC)^{-1},$$

8 $$I - QP = (I + CP)^{-1}.$$

Hence the global input-output and input-error transfer functions of Σ, viz., H_{yu} resp. H_{eu}, have the following form:

9[a] $$H_{yu} = \begin{bmatrix} H_{y_1u_1} & H_{y_1u_2} \\ H_{y_2u_1} & H_{y_2u_2} \end{bmatrix} = \begin{bmatrix} C(I + PC)^{-1} & -CP(I + CP)^{-1} \\ PC(I + PC)^{-1} & P(I + CP)^{-1} \end{bmatrix}$$

9[b] $$= \begin{bmatrix} Q & -QP \\ PQ & P(I - QP) \end{bmatrix},$$

10[a] $$H_{eu} = \begin{bmatrix} H_{e_1u_1} & H_{e_1u_2} \\ H_{e_2u_1} & H_{e_2u_2} \end{bmatrix} = \left[\begin{array}{c|c} (I + PC)^{-1} & -P(I + CP)^{-1} \\ \hline C(I + PC)^{-1} & (I + CP)^{-1} \end{array}\right]$$

10[b] $$= \left[\begin{array}{c|c} I - PQ & -P(I - QP) \\ \hline Q & I - QP \end{array}\right].$$

Moreover, H_{yu} and H_{eu} are related by

11[a] $$H_{eu} = I + FH_{yu}$$

11[b] $$H_{yu} = F - FH_{eu},$$

where

11[c] $$F = \begin{bmatrix} 0 & -I \\ I & 0 \end{bmatrix}; \text{ hence } F^{-1} = -F.$$

For design purposes note the following relations:

12[a] $$H_{y_2d_o} = -H_{e_1d_o} = H_{e_1u_1} = (I + PC)^{-1} = I - PQ,$$

12[b] $$-H_{y_1d_o} = -H_{e_2d_o} = H_{e_2u_1} = C(I + PC)^{-1} = Q,$$

13 $$C = Q(I - PQ)^{-1} = (I - QP)^{-1}Q.$$

Equations (9[b]), (10[b]), and (13) show that <u>Q together with P determines all the properties of Σ</u>.

∎

We now have

14 Theorem [Well posedness]. Consider feedback system Σ, shown in Fig. 1, where we assume that

2 $P \in \mathbb{R}_{p,o}(s)^{n_o \times n_i}$

15 $C \in \mathbb{R}(s)^{n_i \times n_o}$

and

4 $Q := C(I + PC)^{-1} \in \mathbb{R}(s)^{n_i \times n_o}$.

U.t.c.

16 (a) $Q \in E(\mathbb{R}_p(s)) \Leftrightarrow H_{yu}$ and $H_{eu} \in E(\mathbb{R}_p(s))$

17 (b) $Q \in E(\mathbb{R}_p(s)) \Leftrightarrow C \in E(\mathbb{R}_p(s))$

18 (c) $Q \in E(\mathbb{R}_{p,o}(s)) \Leftrightarrow C \in E(\mathbb{R}_{p,o}(s))$.

19 Comments. (a) By eqn. (9), (10), and (12), H_{eu} and $H_{yu} \in E(\mathbb{R}_p(s))$ if and only if all closed-loop transfer functions of feedback system Σ are proper; hence we define feedback system Σ to be well-posed iff all closed-loop transfer functions are well defined and proper. Consequently by (16), once the plant is strictly proper, (see (2)), assumption (3^c) is a necessary and sufficient condition for the well-posedness of Σ.
(b) From (17) it follows also that, with a strictly proper plant, assumption (3) is necessary and sufficient to guarantee a proper compensator: a usual requirement in view of the presence of noise.
(c) From (18) it follows that, with a strictly proper plant, a strictly proper compensator is guaranteed iff Q is strictly proper.

20 Proof of Theorem 14. $\mathbb{R}_p(s)$ is a ring. Hence, since $P \in E(\mathbb{R}_{p,o}(s))$ by assumption (2), it follows from (9)-(10) that (16) holds. Furthermore, since $P \in E(\mathbb{R}_{p,o}(s))$, it follows that (17) and (18) are consequences of (3^b) and (13). ∎

Consider now $R(0)$ the Euclidean ring of exp. stable transfer functions (see Fact 2.2.6), given by

25 $$\mathcal{R}(0) := \{f \in \mathbb{R}_p(s) : f \text{ is analytic in } \mathbb{C}_+\}$$

and let $\mathcal{R}_o(o)$ denote the subring of strictly proper exp. stable transfer functions given by

26 $$\mathcal{R}_o(0) := \{f \in \mathbb{R}_{p,o}(s) : f \text{ is analytic in } \mathbb{C}_+\}.$$

27 Discussion of exponential stability. Under assumptions (2)-(3), feedback system Σ, shown in Fig. 1, is an interconnected system as in Sec. 4.2, satisfying Assumptions IS (see (4.2.3)) and WP (see (4.2.19)-(4.2.21)). (Note that assumption WP holds by Theorem 14.) Let us now assume throughout this chapter that the plant P and compensator C have underlying PMDs which are well formed and have no unstable hidden modes as in Assumption PMD (see (4.2.25). Then, in view of Theorem 4.2.84, the following definition makes sense.

28 Definition. Under assumptions (2)-(3), feedback system Σ, shown in Fig. 1, is said to be exp. stable iff the global input-output transfer function $H_{yu} \in E(\mathcal{R}(0))$.

29 Comment. Note that in view of (11)-(12), Σ is exp. stable iff all closed-loop transfer functions of $\Sigma \in E(\mathcal{R}(0))$.

We have now

30 Theorem [Global Q-parametrization of all stable feedback systems Σ]. Consider feedback system Σ, shown in Fig. 1, under assumptions (2)-(3). U.t.c., if

31 $$P \in E(\mathcal{R}_o(0)),$$

then

32 (a) $$Q \in E(\mathcal{R}(0)) \Leftrightarrow \Sigma \text{ is exp. stable and } C \in E(\mathbb{R}_p(s))$$

33 (b) $$Q \in E(\mathcal{R}_o(0)) \Leftrightarrow \Sigma \text{ is exp. stable and } C \in E(\mathbb{R}_{p,o}(s)).$$ ∎

34 Comment. Theorem 30 teaches us that, given an exp. stable and strictly proper plant P, an exp. stable (and strictly proper) parameter transfer function Q parametrizes all exp. stable feedback systems Σ with a proper

(resp. strictly proper) compensator C. Moreover, eqns. (9)-(13) show that Q determines everything. The design task is therefore reduced to choosing an appropriate $Q \in E(R(0))$ (resp. $Q \in E(R_o(0))$) that satisfies the design specifications required in addition to closed-loop stability and properness of C (see below).

Proof of Theorem 30. Note that $R(0)$ is a ring and that Σ is exp. stable iff $H_{yu} \in E(R(0))$ (by Definition 28). Note also that $R_o(0)$ is a subring of $R(0)$. Hence in view of (31), (32) is a consequence of (9^b), the closure properties of $R(0)$, and Theorem 14. A similar argument establishes (33). ∎

35 Q-parametrization design trade-off. Consider feedback system Σ, shown in Fig. 1, and assume that $P \in E(R_o(0))$ and $Q \in E(R(0))$ (whence, by Theorem 30, Σ is exp. stable and $C = Q(I - PQ)^{-1} \in E(\mathbb{R}_p(s))$). Then from (9)-(10) we have that

(a) the I/O map $H_{y_2u_1}$ of Σ satisfies

36 $$H_{y_2u_1} = PQ;$$

(b) the input to plant-input map $H_{e_2u_1}$ of Σ satisfies

37 $$H_{e_2u_1} = Q;$$

(c) the sensitivity transfer function of Σ (see Sec. 1.3) satisfies

38 $$(I + PC)^{-1} = H_{e_1u_1} = H_{y_2d_o} = I - PQ.$$

Formulas (36)-(38) are important because they display the basic design trade-off:

(1) From (36), (38), and (1.3.52) we see that if

39 $$\|(I - PQ)(j\omega)\| << 1 \quad \forall\omega \in \Omega,$$

where Ω is the frequency band of interest, then, over Ω, (a) the I/O map $H_{y_2u_1}$ is approximately I, (b) we obtain desensitization of the closed-loop system output y_2 due to disturbances d_o applied at the output of the plant, and (c) we obtain desensitization of the I/O map $H_{y_2u_1}$ due to plant variations.

(2) On the other hand, by (37), to prevent saturation at the input of the plant one must also have

40 $$\|Q(j\omega)\| \le K_1 \quad \forall\omega \in \Omega$$

for some given $K_1 > 0$. ■

Although (39)-(40) exhibit the main design trade-off, some other limitations may be present due to the fact that the plant has $\mathbb{C}_+$-zeros.

45 Theorem. [Limitations on the I/O map]. Consider feedback system Σ, shown in Fig. 1, where $P \in E(R_o(0))$. Then any achievable I/O map $H_{y_2u_1}$ of an exp. stable feedback system Σ with a proper compensator C is of the form

46 $$H_{y_2u_1} = PQ, \quad Q \in E(R(0)).$$

Hence

47 $$Z[P] \cap \mathbb{C}_+ \subset Z[H_{y_2u_1}] \cap \mathbb{C}_+.$$

48 Comment. Note that due to (46) every $\mathbb{C}_+$-plant zero is a $\mathbb{C}_+$-zero of the I/O map. Note also that due to (46), since $H_{y_2u_1}$ and P are strictly proper and Q is proper, "as $s \to \infty$, $H_{y_2u_1}$ tends to zero at least as fast as P."

50 Proof of Theorem 45. That any achievable I/O map is the form (46) is a consequence of Theorem 30 (exercise). In order to prove (47), let $(D_{p\ell}, N_{p\ell})$ be a ℓ.c.f. of P and let (N_{qr}, D_{qr}) be a r.c.f. of Q. Observe now that, since $P \in E(R_o(0))$ and $Q \in E(R(0))$, $\det[D_{p\ell}(s)] \neq 0 \;\; \forall s \in \mathbb{C}_+$, and $\det[D_{qr}(s)] \neq 0 \;\; \forall s \in \mathbb{C}_+$. Hence $H_{y_2u_1} = PQ = D_p^{-1}N_p \, N_{qr}D_{qr}^{-1}$ is a left-right fraction of $H_{y_2u_1}$ with no unstable hidden modes. As a consequence (exercise), $z \in Z[P] \cap \mathbb{C}_+ \;\Rightarrow\; z \in Z[H_{y_2u_1}] \cap \mathbb{C}_+$. ■

51 Exercise. For the system Σ of Fig. 1, let (2) and (3) hold and let Σ be exp. stable. Let $n_i \geq n_o$ and normal rank P = normal rank C = n_o. Using int. pr. coprime fractions, we have

$$H_{y_2u_1} = N_{pr}[D_{c\ell}D_{pr} + N_{c\ell}N_{pr}]^{-1}N_{c\ell}.$$

Prove the following statements:

(i) If $z \in \mathbb{C}_+$, then $z \in Z[P] \Rightarrow z \in Z[H_{y_2u_1}]$ and $z \in Z[C] \Rightarrow z \in Z[H_{y_2u_1}]$.

(ii) $\forall s \in \mathbb{C}_+$, $R[N_{pr}(s)] \supset R[H_{y_2u_1}(s)]$. Give a dynamical interpretation when $z \in Z[P] \cap \mathbb{C}_+$.

(iii) $\forall z$ s.t. $\det[D_{c\ell}D_{pr} + N_{c\ell}N_{pr}](z) \neq 0$, $z \in Z[P] \Rightarrow z \in Z[H_{y_2u_1}]$.

52 Exercise. For the system Σ of Fig. 1, let (2) and (3) hold, where P and C are as in Exercise 51; furthermore, insert in the feedback path a transfer function $F(s) \in \mathbb{R}_p(s)^{n_o \times n_o}$ with int. pr. r.c.f. (N_{fr}, D_{fr}). Assume that the resulting feedback system is exp. stable. Prove the following statements:

(i) The I/O map $u_1 \mapsto y_2$ is described by the PMD

$$\mathcal{D}_{y_2u_1} = \left(\begin{bmatrix} D_{c\ell}D_{pr} & N_{c\ell}N_{fr} \\ N_{pr} & -D_{fr} \end{bmatrix}, \begin{bmatrix} N_{c\ell} \\ 0 \end{bmatrix}, [N_{pr} \,\vdots\, 0], 0 \right).$$

(ii) If $z \in \mathbb{C}_+$, $z \in Z[P] \Rightarrow z \in Z[H_{y_2u_1}]$.

(iii) $\forall s \in \mathbb{C}_+$, $R[N_{pr}(s)] \supset R[H_{y_2u_1}(s)]$.

53 Exercise. Consider the system Σ_2 shown in Fig. 2. The plant $P \in \mathbb{R}_{p,o}(s)^{n_o \times n_i}$ has an int. pr. r.c.f. (N_{pr}, D_{pr}). The two-input one-output compensator, is specified by two transfer functions and their int. pr. left fractions

$$G_{y_1v_1} = D_{c\ell}^{-1}N_{\pi} \quad \text{and} \quad G_{y_1e_1} = D_{c\ell}^{-1}N_{c\ell},$$

where

$$\text{rk}[D_{c\ell} \,\vdots\, N_{c\ell} \,\vdots\, N_{\pi}](s) = n_o \quad \forall s \in \mathbb{C}.$$

Assume that Σ_2 is exp. stable. Prove the following.

(i) Write $D_k := D_{c\ell}D_{pr} + N_{c\ell}N_{pr}$; then

$$H_{y_2u_1} = N_{pr}D_k^{-1}N_{\pi}.$$

(ii) $\forall z \in \mathbb{C}_+$, $z \in Z[P] \Rightarrow z \in Z[H_{y_2u_1}]$.

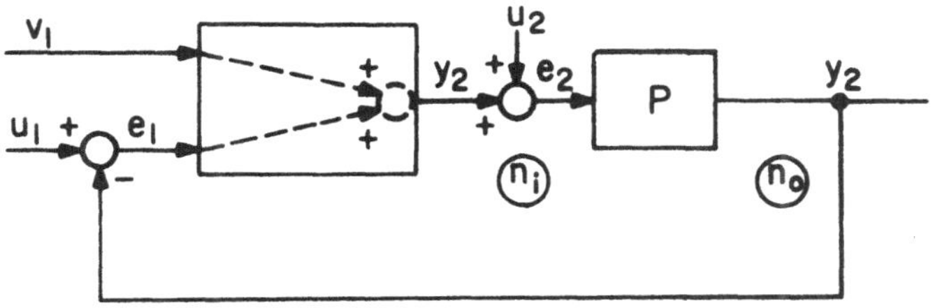

Fig. 2. System Σ_2.

(iii) $\forall s \in \mathbb{C}_+$, $R[N_{pr}(s)] \supset R[H_{y_2u_1}(s)]$.

In (4.4.3) we have seen for system Σ of Fig. 1 that, if P has $\mathbb{C}_+$-zeros and if the loop gain is multiplied by some factor $k \nearrow \infty$, then each $\mathbb{C}_+$-zero of the plant is approached arbitrarily closely by a closed-loop eigenvalue. In view of Sec. 1.3 this suggests that the achievable desensitization in (39) is bounded because of these $\mathbb{C}_+$-zeros. We propose to consider this problem from a slightly different point of view.

Bound on achievable desensitization due to plant $\mathbb{C}_+$-zeros. Let the plant be specified by

60 $$P \in R_o(0)^{n_o \times n_i} \text{ and normal rank of } P = n_o .$$

Consider the system Σ, shown in Fig. 1, with

61 $$Q \in R_o(0)^{n_i \times n_o}.$$

Hence by Theorem 30, Σ is exp. stable. Note now that Fig. 1 is Fig. 1.3.1 with F = I: the sensitivity transfer function of Σ is (using $Q := C(I + PC)^{-1}$)

38 $$(I + PC)^{-1} = H_{e_1u_1} = H_{y_2d_o} = I - PQ,$$

and, from eqn. (1.3.32), if P is slightly perturbed to $\tilde{P}$, then, with higher-order terms in $(\tilde{P} - P)$ neglected,

62 $$\Delta H_{y_2u_1} = (I + PC)^{-1}\Delta H^o_{y_2u_1} = (I - PQ)\Delta H^o_{y_2u_1} .$$

Thus to achieve desensitization we should keep $I - PQ$ as small as possible over the frequency band Ω of interest (see (39)).

63 Remark. Any realistic model requires a strictly proper P and at least a proper C (hence $Q = C(I + PC)^{-1}$ is proper); consequently, as $\omega \to \infty$ $(I - PQ)(j\omega) \to I$. Hence we can keep $I - PQ$ small only over some finite band Ω of frequencies.

Following [Zam.1] we use a weighting function

64 $$w(s) \in R_0(0).$$

Typically, $|w(j\omega)| \simeq 1$, $\forall \omega \in \Omega$, where Ω is the band of interest, and drops sharply to zero outside (e.g., Butterworth functions, [Tem.1]). The design must be closed-loop exp. stable ($\Leftrightarrow Q \in E(R(0))$) and must give the largest amount of desensitization: hence the minimum problem

65 $$\inf\{\sup_{\omega \geq 0} \|w(j\omega)(I - PQ)(j\omega)\|_\infty : Q \in R(0)^{n_i \times n_o}\} =: m(P).$$

We use ℓ_∞-induced norms for technical reasons. Note that for $Q = 0$, $m(P) = 1$.

Let the plant P in (60) have $z_1, z_2, \cdots, z_k$ as $\mathbb{C}_+$-zeros: hence

66 $$\text{rk}[P(z_i)] < n_o \quad \forall i \in \underline{k}$$

with $\text{rk}[P(z)] = n_o$ elsewhere in C_+.

A lower bound on achievable desensitization is given by the following.

70 Theorem [Lower bound on desensitization]. For the plant P satisfying (60) and (66), with the weighting function subject to (64), the minimum of problem (65) satisfies

71 $$m(P) \geq |w(z_i)|, \quad \forall i \in \underline{k}.$$

72 Comment. The lower bound in (71) depends on the location of the z_i's in $\mathbb{C}_+$. For example, for $w(s) = (1+s)^{-n}$, if z_i is definitely outside the band of interest (e.g., $|z_i| \gg 1$), the corresponding $|w(z_i)|$ is very small. However, if $|z_i| < 1$, $|w(z_i)|$ may be close to 1.

73 Proof of Theorem 70. $\forall i \in \underline{k}$, $\forall Q \in R(0)^{n_i \times n_o}$, $P(z_i)\, Q(z_i)$ is singular; hence choose $\eta_i \in \mathbb{C}^{n_o}$ with $\|\eta_i\| = 1$ (here $\|\cdot\|$ denotes ℓ_∞-norms), such that

74 $$P(z_i)Q(z_i)\eta_i = \theta .$$

Let

75 $$f(s, \eta) := w(s)[I - P(s)Q(s)]\eta$$

with $f : \mathbb{C}_+ \times \mathbb{C}^{n_o} \to \mathbb{C}^{n_o}$; note that, $\forall \eta$, $s \mapsto f(s,\eta)$ is analytic in $\mathbb{C}_+$. Now, by (65),

76 $$m(P) = \inf_Q \sup_\omega \sup_\eta \|f(j\omega, \eta)\| ,$$

where $Q \in R(0)^{n_i \times n_o}$, $\omega \geq 0$, $\|\eta\| = 1$. Hence

77 $$m(P) \geq \inf_Q \sup_i \sup_\omega \|f(j\omega, \eta_i)\| .$$

By the maximum modulus theorem and (74), we have considering the ℓth component:

78 $$\sup_\omega |f_\ell(j\omega, \eta_i)| \geq |f_\ell(z_i, \eta_i)| = |w(z_i)\, \eta_{i\ell}| ,$$

where the subscript ℓ denotes the ℓth component of the corresponding vector. Now, by (78) and $\|\eta_i\| = 1$,

79 $$\sup_\omega \|f(j\omega, \eta_i)\| \geq \max_\ell |w(z_i)\, \eta_{i\ell}| = |w(z_i)| .$$

Since the dependence on Q has disappeared from (79), (77) becomes

80 $$m(P) \geq \sup_i |w(z_i)| .$$ ■

8.3. Q-Design Algorithm for Decoupling by Feedback

To illustrate the power of the results of Sec. 8.2, we shall consider a plant $P \in R_o(0)^{n \times n}$ and display a procedure for obtaining a strictly proper compensator C s.t.

(i) the closed-loop feedback system Σ of Fig. 8.2.1 is exp. stable;

(ii) the I/O map $H_{y_2u_1}$ is <u>decoupled</u> and <u>strictly proper</u>;

(iii) in each diagonal element, the poles and zeros (in addition to the $\mathbb{C}_+$-zeros of P) can be specified by the designer.

1 <u>Algorithm</u> [Decoupled I/O map] [Des. 5].

<u>Data</u>: $P \in R_0(0)^{n\times n}$.

<u>Step 1</u>. Obtain an int. pr. r.c.f. (N_{pr}, D_{pr}) of P.

<u>Step 2</u>. <u>If</u> $Z[P] \cap \mathbb{C}_+ = \phi$, pick $n_{j+}(s) \equiv 1 \quad \forall j \in \underline{n}$, and go to step 4; <u>otherwise</u>, go to step 3.

<u>Step 3</u>. Calculate $[\gamma_{ij}]_{i,j\in\underline{n}} =: N_{pr}^{-1}$

and

choose n polynomials $n_{j+}(s)$ of least degree s.t.

2 $\forall i \in \underline{n} \quad \gamma_{ij}(s)n_{j+}(s)$ is analytic in $\mathbb{C}_+$.

<u>Comment</u>: For each j, the polynomial n_{j+} will cancel all the $\mathbb{C}_+$-poles of the jth column of N_{pr}^{-1}. Hence $P^{-1}\cdot\text{diag}[n_{j+}]_{j=1}^{n}$ will be analytic in $\mathbb{C}_+$: we have canceled all $\mathbb{C}_+$-poles of P^{-1}.

<u>Step 4</u>. Choose $\tilde{n}_j(s)$ and $d_j(s)$ $\forall j \in \underline{n}$ in

3 $$H_{y_2u_1}(s) := \text{diag}\left[\frac{n_{j+}(s)\tilde{n}_j(s)}{d_j(s)}\right]_{j=1}^{n}$$

s.t. for every j

4 (i) $Z[d_j] \subset \mathring{\mathbb{C}}_-$,

5 (ii) the polynomial $\tilde{n}_j$ is chosen freely,

6 (iii) $\partial[d_j] > \partial[n_{j+}] + \partial[\tilde{n}_j] + \partial\gamma_j[P^{-1}]$,

where $\partial\gamma_j[P^{-1}]$ denotes the largest degree difference between the numerator and denominator of any element of column j of P^{-1}.

<u>Comments</u>. (a) From (8.2.9), we have $Q = P^{-1} H_{y_2u_1}$, whence (2)-(4) guarantee that Q has no $\mathbb{C}_+$-poles, and (6) guarantees that Q is strictly proper. As a

consequence, $Q \in E(R_o(0))$ ($\Leftrightarrow \Sigma$, shown in Fig. 8.2.1 with $P \in E(R_o(0))$, is exp. stable (Theorem 8.2.30)).

(b) For obtaining $\partial\gamma_j[P^{-1}]$ the computation of $P^{-1} = D_{pr} N_{pr}^{-1}$ is needed. (Note that P^{-1} is also needed in step 5 below.

Step 5. Calculate the required controller transfer function: let

7 $$n_j(s) := n_{j+}(s)\, \tilde{n}_j(s) \quad \forall j \in \underline{n};$$

then

8 $$C(s) := P(s)^{-1} \operatorname{diag}\left[\frac{n_j(s)}{d_j(s) - n_j(s)}\right]_{j=1}^{n}.$$

Comment. Equation (8) follows from (8.2.13) and (8.2.9). We have $C = Q[I - PQ]^{-1}$, $Q = P^{-1}H_{y_2u_1}$. From (3) and (7) we have $H_{y_2u_1} = \operatorname{diag}[n_j/d_j]$. Hence $C = P^{-1}H_{y_2u_1}(I - H_{y_2u_1})^{-1} = P^{-1} \operatorname{diag}[n_j/d_j] \operatorname{diag}[d_j/(d_j-n_j)]$ $= P^{-1} \operatorname{diag}[n_j/(d_j-n_j)]$.

9 Example. Consider

$$P(s) := \frac{1}{(s+2)^2(s+3)} \left[\begin{array}{c|c} 3s + 8 & 2s^2 + 6s + 2 \\ \hline s^2 + 6s + 2 & 3s^2 + 7s + 8 \end{array}\right] \in R_o(0)^{2\times 2},$$

which has a right coprime factorization

$$P(s) = N_{pr}(s)\, D_{pr}(s)^{-1} = \begin{bmatrix} 3 & 2 \\ s + 2 & 3 \end{bmatrix} \begin{bmatrix} s^2 + 3s + 4 & 2 \\ 2 & s + 4 \end{bmatrix}^{-1}.$$

Since $Z[P] = Z[\det N_{pr}] = \{2.5\} \subset \mathbb{C}_+$, we must go through step 3. Now since

$$N_{pr}(s)^{-1} = \begin{bmatrix} \dfrac{-1.5}{s-2.5} & \dfrac{1}{s-2.5} \\[2ex] \dfrac{0.5(s+2)}{s-2.5} & \dfrac{-1.5}{s-2.5} \end{bmatrix},$$

we choose $n_{1+}(s) = n_{2+}(s) = s-2.5$.

So using $Q = P^{-1}H_{y_2u_1}$, with $H_{y_2u_1}$ given by (3), we have

$$Q(s) = -0.5\begin{bmatrix} \frac{(3s^2+7s+8)\tilde{n}_1(s)}{d_1(s)} & \frac{-(2s^2+6s+2)\tilde{n}_2(s)}{d_2(s)} \\ \frac{-(s^2+6s+2)\tilde{n}_1(s)}{d_1(s)} & \frac{(3s+8)\tilde{n}_2(s)}{d_2(s)} \end{bmatrix}.$$

To guarantee $Q \in \mathcal{R}_o(0)^{2\times 2}$, we choose $\tilde{n}_1(s) = \tilde{n}_2(s) = 1$ and $d_1(s)$, $d_2(s)$ s.t.

(i) $\partial[d_1] \geq 3$, $\partial[d_2] \geq 3$,

(ii) $Z[d_j] \subset \mathring{\mathbb{C}}_-$ for $j = 1, 2$.

Then the resulting I/O map $H_{y_2u_1}$ is given by

10 $$H_{y_2u_1}(s) = \text{diag}\left[\frac{s-2.5}{d_1(s)}, \frac{s-2.5}{d_2(s)}\right],$$

and the resulting compensator is given by

11 $$C(s) = \begin{bmatrix} \frac{-0.5(3s^2+7s+8)}{d_1(s)-(s-2.5)} & \frac{s^2+3s+1}{d_2(s)-(s-2.5)} \\ \frac{0.5(s^2+6s+2)}{d_1(s)-(s-2.5)} & \frac{-0.5(3s+8)}{d_2(s)-(s-2.5)} \end{bmatrix}.$$

Note finally that if in (10) $H_{y_2u_1}(s)\big|_{s=0} = I$, feedback system Σ will asymptotically track any step input. ∎

12 Exercise. Consider Example 9 and construct a compensator C such that feedback system Σ is exp. stable, tracks asymptotically any ramp, and $H_{y_2u_1}$ is diagonal. (See Exercise 7.2.71.) ∎

8.4. Two-Step Compensation Theorem for Unstable Plants

In this section we explore the following questions: (i) if the plant P is exp. unstable, may we first stabilize it and then, for the purpose of design, find a global parametrization of all stable feedback systems similar to Theorem 8.2.30? (ii) If so, is any achievable I/O map $H_{y_2u_1}$ achievable by the two-step scheme described in (i)?

Let us first specify our assumptions

1 Assumption. The plant is specified by its transfer function $P_o \in \mathbb{R}_{p,o}(s)^{n_o \times n_i}$, which is generated by a well-formed plant PMD $\mathcal{D}_{P_o}$ with no unstable hidden modes and s.t.

$$P_o \notin E(R_o(0)).$$

2 Assumption. The plant P_o can be stabilized by an exp. stable local feedback, equiv. there exists a feedback $F_0 \in R(0)^{n_i \times n_o}$, generated by a well-formed PMD $\mathcal{D}_{f_o}$ with no unstable hidden modes, which results in an exp. stable single-loop feedback system ${}^1S(F_0, P_o)$ (see Fig. 1), with the stabilized plant I/O map given by

3 $$H_{y_2'e_2''} =: P_1 = P_o(I + F_0P_o)^{-1} = (I + P_oF_o)^{-1}P_o.$$

4 Comments. (a) The purpose of Assumption 1 is to guarantee the existence of an int. pr. compensator $C \in \mathbb{R}_p(s)^{n_i \times n_o}$ s.t. the single-loop feedback system ${}^1S(P_o, C)$, shown in Fig. 2 is exp. stable.

For example, $P_o \in \mathbb{R}_{p,o}(s)^{n_o \times n_i}$ is generated by the well-formed PMD $\mathcal{D}_{P_o} = [D_{P_o r}, I, N_{P_o r}, 0]$, where $(N_{P_o r}, D_{P_o r})$ is an int. pr. r.c.f. of P_o with $D_{P_o r}$ column reduced; by the method of Sec. 6.2, we can then find an int. pr. compensator $C \in \mathbb{R}_p(s)^{n_i \times n_o}$ s.t. ${}^1S(P_o, C)$ is exp. stable.

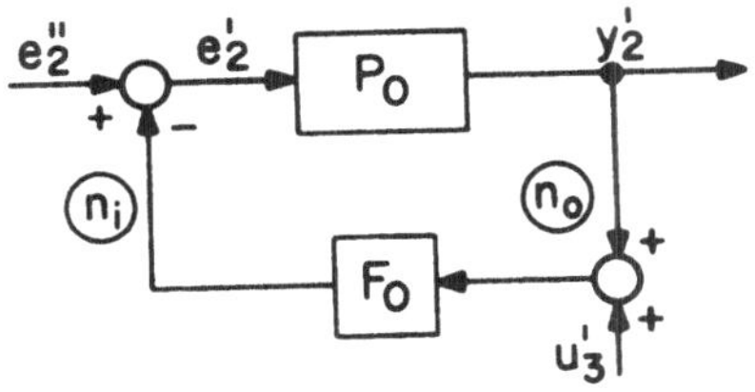

Fig. 1. The system ${}^1S(F_0, P_o)$.

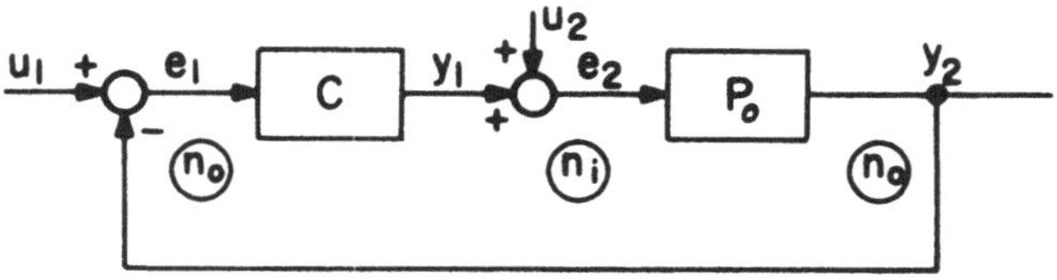

Fig. 2. The system ${}^1S(P_o, C)$.

(b) The notation used in Fig. 1 for system ${}^1S(F_o, P_o)$ will become clear in the sequel.
(c) Comparing Fig. 1 and Fig. 2, we see that systems ${}^1S(F_o, P_o)$ and ${}^1S(P_o, C)$ are similar if we interchange F_o with P_o and P_o with C. Hence in view of definition (8.2.3^b) the parameter of ${}^1S(F_o, P_o)$ reads $P_o(I + F_oP_o)^{-1} = P_1$. Consequently, by an analysis similar to the one done for Theorem 8.2.30, we have

5 Fact. Consider the single-loop feedback system ${}^1S(F_o, P_o)$, shown in Fig. 1, where P_o satisfies Assumption 1 and $F_o \in E(R(0))$ is as in Assumption 2.
U.t.c.

6 $${}^1S(F_o, P_o) \text{ is exp. stable}$$

$\Leftrightarrow$

the transformed-plant I/O map $P_1 : e_2'' \mapsto y_2'$ satisfies

7 $$P_1 := P_o(I + F_oP_o)^{-1} \in E(R_o(0)).$$ ∎

8 Comments. (a) Note that system ${}^1S(F_o, P_o)$ (see Fig. 1) reduces to P_1 if $u_3' = \theta$ and that by Fact 5 the exp. stability of P_1 is equivalent to the exp. stability of ${}^1S(F_o, P_o)$. Although subsystem P_1 is simpler than system ${}^1S(F_o, P_o)$, the additional external input u_3' provides an additional degree of freedom which enables us to realize all achievable I/O maps $H_{y_2u_1}$ of ${}^1S(P_o, C)$ (see Fig. 2) by the two-loop feedback system studied below (see Exercise 24).

(b) The Assumption 2 that P_0, given by Assumption 1, can be stabilized by an exp. stable feedback $F_0 \in R(0)^{n_i \times n_o}$ is not very restrictive. Indeed [You.1], this is always possible if there exists <u>no</u> finite point z on the nonnegative real axis in $\mathbb{C}_+$ s.t. $P_0(z) = 0$. If P_0 is zero at distinct finite points $z_1, z_2, \cdots, z_\ell$ on the nonnegative real axis in $\mathbb{C}_+$, let ν_i denote the number of $\mathbb{C}_+$-poles of $P_0 \in \mathbb{R}_{p,o}(s)^{n_o \times n_i}$ on the nonnegative real axis to the right of z_i, counting multiplicities according to their Mcmillan degrees, (i.e., their maximal order as a pole of any minor of any order of P_0). Then P_0 is stabilizable by a stable F_0 if and only if the integers ν_i are <u>all even</u>, [This follows by (a) $P_0(\infty) = 0$ and (b) [You.1, Th. 2, Corr. 1]].

We shall now compensate the exp. stable system ${}^1S(F_0, P_0)$ by $C - F_0$, thus giving the two-loop feedback system ${}^3S(P_0, F_0, C - F_0)$, shown in Fig. 3.

9 <u>Forward Comment</u>. Roughly speaking, in Theorem 18 we shall show (a) that the existence of a proper compensator C resulting in an exp. stable system ${}^1S(P_0, C)$ (Fig. 2) is <u>equivalent</u> to the existence of the proper compensator $C - F_0$ s.t. ${}^3S(P_0, F_0, C - F_0)$ is exp. stable, and (b) that <u>all</u> stable systems ${}^3S$ can be globally <u>parametrized</u> by a parameter $Q \in E(R(0))$ s.t. $C - F_0 = Q(I - P_1Q)^{-1}$, where P_1 is the stabilized plant (3) (whence by (a) <u>all</u> stable systems ${}^1S(P_0, C)$ can be parametrized by an exp. stable parameter Q as in

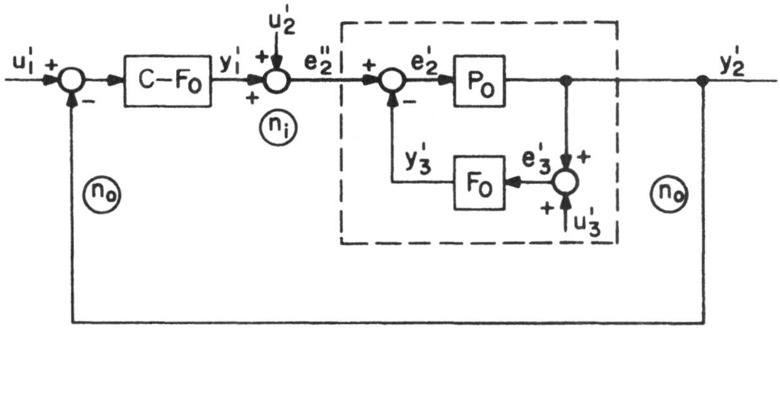

Fig. 3. The system ${}^3S(P_0, F_0, C - F_0)$.

Theorem 8.2.30). Moreover, in Exercise 24 below we shall show that any I/O map $H_{y_2u_1}$, achievable by an exp. stable system ${}^1S(P_o, C)$, can be achieved as the I/O map $H_{y_2'u_1'}$ of the corresponding exp. stable system ${}^3S(P_o, F_o, C - F_o)$ by setting $u_3' = -u_1'$. <u>Consequently, from an algebraic point of view the two-step compensation scheme</u> ${}^3S(P_o, F_o, C - F_o)$ <u>can do whatever the one-step scheme</u> ${}^1S(P_o, C)$ <u>can do</u>. <u>No design opportunity is lost</u>.

10 <u>Analysis</u>. Consider systems ${}^1S(P_o, C)$ and ${}^3S(P_o, F_o, C - F_o)$, shown in Figs. 2 and 3, under Assumptions 1 and 2. Using Laplace transforms, and merging the two middle summing nodes in Fig. 3 into one node, the system equations for ${}^1S$ resp. ${}^3S$ are

11
$$\underbrace{\begin{bmatrix} I & P_o \\ -C & I \end{bmatrix}}_{=:\,M}\begin{bmatrix} e_1 \\ e_2 \end{bmatrix} = \begin{bmatrix} u_1 \\ u_2 \end{bmatrix}, \text{ resp. } \underbrace{\begin{bmatrix} I & P_o & 0 \\ -(C-F_o) & I & F_o \\ 0 & -P_o & I \end{bmatrix}}_{=:\,M'}\begin{bmatrix} e_1' \\ e_2' \\ e_3' \end{bmatrix} = \begin{bmatrix} u_1' \\ u_2' \\ u_3' \end{bmatrix},$$

12
$$\begin{bmatrix} e_1 \\ e_2 \end{bmatrix} = \begin{bmatrix} u_1 \\ u_2 \end{bmatrix} + \underbrace{\begin{bmatrix} 0 & -I \\ I & 0 \end{bmatrix}}_{=:\,F}\begin{bmatrix} y_1 \\ y_2 \end{bmatrix}, \text{ resp. } \begin{bmatrix} e_1' \\ e_2' \\ e_3' \end{bmatrix} = \begin{bmatrix} u_1' \\ u_2' \\ u_3' \end{bmatrix}$$
$$+ \underbrace{\begin{bmatrix} 0 & -I & 0 \\ I & 0 & -I \\ 0 & I & 0 \end{bmatrix}}_{=:\,F'}\begin{bmatrix} y_1' \\ y_2' \\ y_3' \end{bmatrix}.$$

Let

$$n := n_i + n_o, \qquad \text{resp. } n' := n_i + n_o + n_o.$$

Note that F and F' are <u>constant</u>, that F is <u>nonsingular</u> and $F_o \in \mathcal{R}(0)^{n_i \times n_o}$; consequently, using (11)-(12) and Fig. 3

13 $^1S(P_o, C)$ is exp. stable $\Leftrightarrow$ $H_{yu} \in R(0)^{n\times n}$ $\Leftrightarrow$ $H_{eu} \in R(0)^{n\times n}$

$\Leftrightarrow$ $M^{-1} \in R(0)^{n\times n}$,

and

14 $^3S(P_o, F_o, C - F_o)$ is exp. stable $\Leftrightarrow$ $H_{y'u'} \in R(0)^{n'\times n'}$

$\Leftrightarrow$ $H_{e'u'} \in R(0)^{n'\times n'}$ $\Leftrightarrow$ $(M')^{-1} \in R(0)^{n'\times n'}$.

Note with care that, by (12), $H_{y'u'} \in R(0)^{n'\times n'} \Rightarrow H_{e'u'} \in R(0)^{n'\times n'}$; moreover, the relations $y_2' = u_1' - e_1'$, $y_3' = F_o e_3'$, with $F_o \in R(0)^{n_i\times n_o}$, and $y_1' = e_2' - u_2' + y_3'$ imply that $H_{e'u'} \in R(0)^{n'\times n'} \Rightarrow H_{y'u'} \in R(0)^{n'\times n'}$ (note that we cannot use (12), since F' is singular).

14 <u>Comment</u>. In the stability analysis above it is assumed that C and $C - F_o$ are generated by well-formed PMDs with no unstable hidden modes, as is the case for P_o and F_o (see Assumptions 1 and 2). In other words, although the analysis is made in the frequency domain in algebraic terms, there is everywhere an implied time-domain interpretation. Moreover, the same applies to the results below.

18 <u>Theorem</u> [Two-step compensation theorem]. Consider the systems $^1S(P_o, C)$ and $^3S(P_o, F_o, C - F_o)$, shown in Figs. 2 and 3, and described by (11)-(12). Let Assumptions 1 and 2 hold.

U.t.c., for any $F_o \in R(0)^{n_i\times n_o}$ satisfying (2),

19 $\exists\ C \in \mathbb{R}_p(s)^{n_i\times n_o}$ s.t. $^1S(P_o, C)$ is exp. stable,

$\Leftrightarrow$

20 $\exists\ C - F_o \in \mathbb{R}_p(s)^{n_i\times n_o}$ s.t. $^3S(P_o, F_o, C - F_o)$ is exp. stable,

$\Leftrightarrow$

21 $\exists\ Q \in R(0)^{n_i\times n_o}$ s.t. $C - F_o := Q(I - P_1 Q)^{-1}$. ∎

22 <u>Remark</u>. Recall Assumptions 1 and 2, Fact 5, and Theorem 8.2.30. Then, with the Q defined in (21), it follows that

$Q \in R(0)^{n_i\times n_o}$ $\Leftrightarrow$ $^1S(P_1, C - F_o)$ is exp. stable, where ${}^{1}S(P_1, C - F_0)$ is obtained from ${}^{3}S$ by setting $u_3' = \theta$ in Fig. 3.

23 Exercise. Use the Q, defined in (21), and the P_1, defined in (3), to prove that

$$H_{y_1'u_1'} = Q; \; H_{y_2'u_1'} = P_1Q = (I + P_oF_o)^{-1} P_oQ;$$

$$H_{e_1'u_1'} = I - P_1Q = I - (I + P_oF_o)^{-1}P_oQ.$$

24 Exercise [Achievable I/O maps]. In system ${}^{3}S(P_o, F_0, C - F_0)$, shown in Fig. 3, set $u_3' = -u_1'$ (equivalently feedforward u_1' through a $-I$ gain to u_3'), and show that for the resulting system, $H_{y_2'u_1'} = H_{y_2u_1}$, where the latter is the I/O map of ${}^{1}S(P_o, C)$.

(Hint: Use (12): set $u_3' = -u_1'$; perform $\rho_3 \leftarrow \rho_3 + \rho_1$; eliminate e_3'; $\cdots\cdot$.)

25 Exercise. Verify that the forward comment (9) is true. Hence the answer to both questions in the beginning of this section is yes.

26 Comment on the exp. stability of F_0. Consider Figs. 2 and 3, the definition of P_1 in (3), Fact 5, and statements (19), (20), and (21) of Theorem 18. It is intuitively clear that exp. stability of ${}^{3}S(P_o, F_0, C - F_0)$ implies the exp. stability of ${}^{1}S(P_o, C)$ and ${}^{1}S(P_1, C - F_0)$ without requiring that F_0 be exp. stable (this fact will be exhibited in the proofs of (20)⇒(19) and (20)⇒(21) below by appropriately restricting ${}^{3}S$). However the proofs of (19)⇒(20) and (21)⇒(20) require that F_0 be exp. stable; to emphasize that this assumption is indispensable we offer the corollary below.

30 Corollary. Let $P_o \in \mathbb{R}_{p,o}(s)^{n_o \times n_i}$. Let $C \in \mathbb{R}_p(s)^{n_i \times n_o}$ be s.t. ${}^{1}S(P_o, C)$, defined by (11) and (12), is exp. stable. Let

31 $F = \{F_0 \in \mathbb{R}_p(s)^{n_i \times n_o}$: F_0 has at least one $\mathbb{C}_+$-pole, say p_1, s.t. p_1 is not a pole of $C\}$.

U.t.c.

32 $\forall F_0 \in F$, ${}^{3}S(P_o, F_0, C - F_0)$ is not exp. stable.

Proof. On eqn. (11) describing ${}^{3}S(P_o, C_o, C - F_0)$, perform $\gamma_1 \leftarrow \gamma_1 - \gamma_3$ and $\rho_3 \leftarrow \rho_3 + \rho_1$ to obtain

33 $$\begin{bmatrix} I & P_o & 0 \\ -C & I & F_0 \\ 0 & 0 & I \end{bmatrix} \begin{bmatrix} e_1' \\ e_2' \\ e_3'' \end{bmatrix} = \begin{bmatrix} u_1' \\ u_2' \\ u_4' \end{bmatrix},$$

where $e_3'' := e_1' + e_3'$ and $u_4' := u_1' + u_3'$, Easy calculations using (33) give

$$H_{e_1'u_4'} = P_o(I + CP_o)^{-1}F_0,$$
$$H_{e_2'u_4'} = -(I + CP_o)^{-1}F_0.$$

Hence using int. pr. coprime fractions, these relations become

34[a] $$H_{e_1'u_4'} = N_{p_or}(D_{c\ell}D_{p_or} + N_{c\ell}N_{p_or})^{-1}\, D_{c\ell}\, D_{f_o\ell}^{-1}\, N_{f_o\ell},$$

34[b] $$H_{e_2'u_4'} = -D_{p_or}(D_{c\ell}D_{p_or} + N_{c\ell}N_{p_or})^{-1}\, D_{c\ell}\, D_{f_o\ell}^{-1}\, N_{f_o\ell}.$$

Now note that, since (N_{p_or}, D_{p_or}) is r.c., by the Bezout identity, there exist polynomial matrices U, V s.t.

$$UN_{p_or} + VD_{p_or} = I,$$

whence (34) implies

35 $$D_{c\ell}^{-1}(D_{c\ell}D_{p_or} + N_{c\ell}N_{p_or})(UH_{e_1'u_4'} - VH_{e_2'u_4'}) = D_{f_o\ell}^{-1}\, N_{f_o\ell} = F_0.$$

Since $F_0 \in F$, by (31), the RHS of (35) has a pole at $p_1 \in \mathbb{C}_+$, while $\det[D_{c\ell}(p_1)] \neq 0$. Moreover, since by assumption ${}^1S(P_o, C)$ is exp. stable, $\det[(D_{c\ell}D_{p_or} + N_{c\ell}N_{p_or})(p_1)] \neq 0$. Assume now, for the purpose of a contradiction that ${}^3S(P_o, F_0, C - F_0)$ is exp. stable. Then, with $u_4' = u_1' + u_3'$, $H_{e_1'u_4'}$ and $H_{e_2'u_4'} \in E(R(0))$. Hence, with U and $V \in E(\mathbb{R}[s])$, and $D_{c\ell}^{-1}(D_{c\ell}D_{p_or} + N_{c\ell}N_{p_or})$ nonsingular at $p_1 \in \mathbb{C}_+$, the LHS of (35) has no pole at $p_1 \in \mathbb{C}_+$, while the RHS has a pole there. As a consequence we obtain a contradiction; thus, $\forall F_0 \in F$, ${}^3S(P_o, F_0, C - F_0)$ is not exp. stable. ∎

37 Remark. In particular, the proof shows that with $u_4' = u_1' + u_3'$, $H_{e_1'u_4'}$, and/or $H_{e_2'u_4'}$ must have a pole at $p_1 \in \mathbb{C}_+$. Hence if $u_1'(s) = u_3'(s)$

$= k(s - \alpha)^{-1}$, where $k \in \mathbb{C}^{n_o}$ and Re $\alpha < 0$, then for some k the responses $e_1'(s)$ and/or $e_2'(s)$ have a pole at $p_1 \in \mathbb{C}_+$ (at least one corresponding time function includes a term $O(\exp p_1 t)$ with $p_1 \in \mathbb{C}_+$).

40 Proof of Theorem 18

(19) ⇒ (20). By assumption ${}^1S(P_o, C)$ is exp. stable, whence, by (13), $M^{-1} \in E(R(0))$. We claim that M', defined in eqn. (11), satisfies $(M')^{-1} \in E(R(0))$, whence, by (14), ${}^3S(P_o, F_0, C - F_0)$ is exp. stable.

On M', perform $\gamma_1 \leftarrow \gamma_1 - \gamma_3$, $\rho_3 \leftarrow \rho_3 + \rho_1$, and $\rho_2 \leftarrow \rho_2 - F_0\rho_3$ to obtain

41 $$M_3 = \begin{bmatrix} I & P_o & 0 \\ -C & I & 0 \\ 0 & 0 & I \end{bmatrix}.$$

Since, by assumption, $F_0 \in E(R(0))$, $M_3 = LM'R$, where L and R are unimodular matrices with elements in $R(0)$. Consequently, by inspection of (41), $M^{-1} \in E(R(0))$ implies $(M')^{-1} \in E(R(0))$.

(20) ⇒ (19). By assumption, ${}^3S(P_o, F_0, C - F_0)$ is exp. stable. In (11), set $u_3' = -u_1'$ and perform $\gamma_1 \leftarrow \gamma_1 - \gamma_3$ and $\rho_3 \leftarrow \rho_3 + \rho_1$. Then, with $e_3'' := e_1' + e_3'$,

$$\begin{bmatrix} I & P_o & 0 \\ -C & I & F_0 \\ 0 & 0 & I \end{bmatrix} \begin{bmatrix} e_1' \\ e_2' \\ e_3'' \end{bmatrix} = \begin{bmatrix} u_1' \\ u_2' \\ \theta \end{bmatrix}.$$

By the third equation $e_3'' = \theta$, whence

44 $$\begin{bmatrix} I & P_o \\ -C & I \end{bmatrix} \begin{bmatrix} e_1' \\ e_2' \end{bmatrix} = \begin{bmatrix} u_1' \\ u_2' \end{bmatrix}.$$

So with $u_3' = -u_1'$, the partial input-error map $(u_1', u_2') \mapsto (e_1', e_2')$ of ${}^3S$ is identical with the input-error map of ${}^1S(P_o, C)$ (see (11) and (12)). Hence, using (13) and (14), the exp. stability of ${}^3S(P_o, F_0, C - F_0)$ implies the exp. stability of ${}^1S(P_o, C)$.

(20) ⇒ (21). If we show that the exp. stability of ${}^3S(P_o, F_0, C - F_0)$ implies the exp. stability of ${}^1S(P_1, C - F_0)$, then by global parametrization Theorem 8.2.30 and Fact 5, assertion (21) follows.

By assumption, for the system ${}^3S(P_o, F_0, C - F_0)$, the maps $(u_1', u_2', u_3') \mapsto (e_1', e_2', e_3')$ and $(u_1', u_2', u_3') \mapsto (y_1', y_2', y_3')$ are exp. stable (equiv. $\in E(R(0))$). Hence with $e_2'' := y_1' + u_2'$ and $u_3' = \theta$ the partial map $(u_1', u_2') \mapsto (e_1', e_2'')$ is exp. stable. Now, refer to Fig. 3 and observe that $e_2'' = y_1' + u_2'$ is an external input to subsystem ${}^1S(F_0, P_o)$, which reduces to $P_1 := P_o(I + F_0P_o)^{-1}$ when $u_3' = \theta$. Hence, with $u_3' = \theta$, the partial map $(u_1', u_2') \mapsto (e_1', e_2'')$ of ${}^3S$ is the input-error map of ${}^1S(P_1, C - F_0)$: since the latter map is exp. stable, it follows that ${}^1S(P_1, C - F_0)$ is exp. stable.

<u>(21) ⇒ (20)</u>

<u>Step 1</u>. $Q \in E(R(0))$ and $C - F_0 = Q(I - P_1 Q)^{-1}$ imply that ${}^1S(P_1, C - F_0)$ is exp. stable by Theorem 8.2.30 and Fact 5. Consequently, with $u_3' = \theta$ (see Fig. 3) and with $e_2' = e_2'' - F_0y_2'$, we conclude that the maps

46 $$\begin{cases} (u_1', u_2') \mapsto (e_1', e_2''),\ (u_1', u_2') \mapsto (y_1', y_2') \text{ and } (u_1', u_2') \mapsto (e_1', e_2') \\ \text{are exp. stable.} \end{cases}$$

Since with $u_3' = \theta$ the eqns. relating (u_1', u_2') to (e_1', e_2') in Fig. 3 are

47 $$\underbrace{\begin{bmatrix} I & P_o \\ -(C - F_0) & I + F_0P_o \end{bmatrix}}_{=: M_4} \begin{bmatrix} e_1' \\ e_2' \end{bmatrix} = \begin{bmatrix} u_1' \\ u_2' \end{bmatrix},$$

we conclude that $M_4^{-1} \in E(R(0))$.

<u>Step 2</u>. Consider ${}^3S(P_o, F_0, C - F_0)$. Then on (11) perform $\rho_2 \leftarrow \rho_2 - F_0\rho_3$, $\rho_3 \leftarrow \rho_3 + \rho_1$, and $\gamma_1 \leftarrow \gamma_1 - \gamma_3$ to obtain

50 $$\underbrace{\begin{bmatrix} I & P_o & 0 \\ -(C - F_0) & I + F_oP_o & 0 \\ 0 & 0 & I \end{bmatrix}}_{=: M_5} \begin{bmatrix} e_1' \\ e_2' \\ e_1' + e_3' \end{bmatrix} = \begin{bmatrix} u_1' \\ u_2' - F_0u_3' \\ u_3' + u_1' \end{bmatrix}.$$

We have used only row operations in the ring $R(0)$, hence

51 $$(M_5)^{-1} \in E(R(0)) \Leftrightarrow (M')^{-1} \in E(R(0)).$$

<u>Step 3</u>. From step 1, (21) $\Rightarrow M_4^{-1} \in E(R(0))$. By inspection of (47) and (50), $M_4^{-1} \in E(R(0)) \Leftrightarrow M_5^{-1} \in E(R(0))$. Consequently, by (51) and (14), assumption (21) implies that ${}^3S(P_0, F_0, C - F_0)$ is exp. stable. ∎

52 <u>Comment</u>. Note that the proofs of (20) $\Rightarrow$ (19) and (20) $\Rightarrow$ (21) did <u>not</u> require that $F_0 \in E(R(0))$.

Epilogue

At the close of this volume it is important to keep in mind the following points:

1 Discrete-time systems. Lack of space prevented us to follow each chapter with an isomorphic chapter of results covering the discrete-time case, [Kuc.1]. It is very useful to keep in mind the following rough correspondences.

Continuous-time case	Discrete-time case
$f : \mathbb{R}_+ \to \mathbb{R}^n$; $f : t \mapsto f(t)$	$f : \mathbb{N} \to \mathbb{R}^n$, $f = (f_i)_{i=0}^{\infty}$
$\hat{f}$ is analytic in some right-half plane: $\{s \mid \mathrm{Re}\ s > \rho\}$	$\tilde{f}(z) := \sum_{0}^{\infty} f_k z^{-k}$ is analytic outside the closed disk $\overline{D(0, \rho_1)}$ centered on 0, with radius ρ_1
$\mathbb{C}_+$	$D(0, 1)^c = \{z \mid \lvert z\rvert \geq 1\}$
$\mathring{\mathbb{C}}_-$	$D(0, 1)$ = open disk = $\{z \mid \lvert z\rvert < 1\}$
$\mathbb{R}_p(s)$ = proper rational fn	$\tilde{f} \in \mathbb{R}_p(z) \Rightarrow \tilde{f}$ is a causal transfer function
$\hat{f} \in \mathcal{R}(0) \Leftrightarrow \hat{f}$ is proper and exp. st. ($\hat{f}$ analytic in $\mathbb{C}_+$ including at ∞).	$\tilde{f} \in \tilde{\mathcal{R}}(0) \Leftrightarrow \tilde{f}$ is causal and exp. stable ($\tilde{f}$ analytic in $D(0, 1)^c$ including at ∞).

2 Distributed-systems. The concepts results and techniques developed in this volume have been extended to a general class of linear time-invariant distributed systems [Cal.1], [Vid.1].

3 Robustness of stability considerations. Our discussion considered exclusively lumped linear time-invariant systems. As engineers it is important to know that it can be shown that small nonlinear perturbations, slow time-variations of the coefficients and small delay effects do not upset the conclusions on stability and tracking [Des.4], [Wil.1], [Vid.2], [Vid.3], [Saf.2].

4 Computer-aided design. The recent surge of developments in computer-aided design, e.g., [Pol.1] suggests the following views:

(a) The first task of theory is to delineate what is possible.

(b) Design algorithms must provide families of conveniently parametrized designs that satisfy gross qualitative features (e.g., stability, properness, etc.).

Then the design objective is expressed by a performance function which is, e.g., to be minimized subject to a number of inequality constraints that reflect important engineering requirements (e.g., upper bound on some transfer functions to guard against saturation, upper bound on $\omega \mapsto \bar{\sigma}[(I + PC)(j\omega)^{-1}]$ over some interval to guarantee some degree of desensitization, etc.). Then the designer uses the CAD facility to search in parameter space for the "best" solution.

5 The field of multivariable feedback control is rapidly moving. This can be seen from the abundant literature in journals such as the IEEE Transactions on Automatic Control, Automatica, and International Journal of Control, e.g., [Per.1], and research reports, e.g., [Fra.1].

Appendix A: Rings and Fields

1 Every engineer has performed some computations

(a) in the following <u>fields</u>: $\mathbb{R}$, $\mathbb{C}$, Q; $\mathbb{R}(s)$, $\mathbb{C}(s)$;

(b) in the following <u>commutative rings</u>:

$\mathbb{Z}$, $\mathbb{R}[s]$, $\mathbb{C}[s]$, $\mathbb{R}_p(s)$, $\mathbb{R}_{p,o}(s)$, $\mathcal{R}(0)$, $\mathcal{R}_o(0)$, $\mathcal{R}_U$; diagonal matrices with elements in a field (e.g., $\mathbb{R}$, $\mathbb{C}$, or $\mathbb{R}(s)$) or in a commutative ring (e.g., $\mathbb{R}[s]$, $\mathbb{C}[s]$);

where in particular

2 $\mathbb{R}_p(s)$, $(\mathbb{R}_{p,o}(s))$:= the ring of proper[§] (strictly proper)[†] rational functions with coefficients in $\mathbb{R}$.

3 $\mathcal{R}(0)$, $(\mathcal{R}_o(0))$:= the subring of elements of $\mathbb{R}_p(s)$, $(\mathbb{R}_{p,o}(s))$, that are analytic in $\mathbb{C}_+$ (i.e., with no poles in $\mathbb{C}_+$)

4 $\mathcal{R}_U$:= the subring of elements of $\mathbb{R}_p(s)$ that are analytic in U: a closed subset of $\mathbb{C}$ symmetric w.r.t. the real axis and which includes $\mathbb{C}_+$.

Elements of $\mathcal{R}(0)$ and $\mathcal{R}_U$ are also called <u>exp. stable</u> resp. <u>U-stable transfer functions</u>.

(c) in the following <u>noncommutative rings</u>:

$\mathbb{R}^{n\times n}$, $\mathbb{C}^{n\times n}$; $\mathbb{R}[s]^{n\times n}$, $\mathbb{C}[s]^{n\times n}$; $\mathbb{R}(s)^{n\times n}$, $\mathbb{C}(s)^{n\times n}$; $\mathbb{R}_p(s)^{n\times n}$, $\mathbb{R}_{p,o}(s)^{n\times n}$; $\mathcal{R}(0)^{n\times n}$, $\mathcal{R}_o(0)^{n\times n}$; $\mathcal{R}_U^{n\times n}$,

where we consider $n \times n$ matrices with elements in $\mathbb{R}$, $\mathbb{C}$, etc. ∎

5 <u>Exercise</u>. For each of the rings and fields above, verify that you already know the operations of <u>addition</u> and <u>multiplication</u>. Identify precisely the element 0 and the element 1, the identity under addition and multiplication, resp.

[§] Bounded at infinity;

[†] (zero at infinity).

6 Definition. Any ring and any field is a set of elements together with two binary operations, an addition + and a multiplication • ; an identity element under addition denoted by 0; and an identity element under multiplication denoted by 1, and they obey the following respective axioms:

Axioms

Ring: $(R, +, \cdot; 0, 1)$	Field: $(\mathbb{F}, +, \cdot; 0, 1)$

Addition is

Associative:	$(\alpha + \beta) + \gamma = \alpha + (\beta + \gamma)$	$\forall \alpha, \beta, \gamma$
Commutative:	$\alpha + \beta = \beta + \alpha$	$\forall \alpha, \beta$
$\exists$ identity 0:	$\alpha + 0 = \alpha$	$\forall \alpha$
$\exists$ inverse:	$\forall \alpha$, $\exists$ element $(-\alpha)$ $\ni$	$\alpha + (-\alpha) = 0$

Multiplication is

Associative: $(\alpha \cdot \beta) \cdot \gamma = \alpha \cdot (\beta \cdot \gamma)$ $\quad \forall \alpha, \beta, \gamma$

Ring	Field
Not necessarily commutative	Commutative $\alpha \cdot \beta = \beta \cdot \alpha$ $\forall \alpha, \beta \in \mathbb{F}$

$\exists$ identity 1: $\alpha \cdot 1 = 1 \cdot \alpha = \alpha$, $\forall \alpha$.

Ring	Field
$\alpha \in R$, $\alpha \neq 0 \not\Rightarrow \alpha^{-1}$ exists	$\exists$ Inverse: $\forall \alpha \neq 0$, $\exists$ inverse $a^{-1} \ni$ $\alpha \cdot (\alpha^{-1}) = (\alpha^{-1}) \cdot \alpha = 1$

Distributive law: $\forall \alpha, \beta, \gamma$

D_L: $\alpha \cdot (\beta + \gamma) = \alpha \cdot \beta + \alpha \cdot \gamma$

D_R: $(\beta + \gamma) \cdot \alpha = \beta \cdot \alpha \quad \gamma \cdot \alpha$

From these axioms follow four important facts.

7 Fact. In any ring and in any field, the identities 0 and 1 are unique. (This is easily shown by contradiction.)

8 Fact. In a ring, the cancellation law does not necessarily hold; more precisely, in a ring

$$\left.\begin{array}{l}\alpha\beta = \alpha\gamma \\ \text{and} \\ \alpha \neq 0\end{array}\right\} \quad \text{do not necessarily imply } \beta = \gamma.$$

Example. Consider the noncommutative ring $\mathbb{R}^{2\times 2}$:

$$\begin{bmatrix} 0 & 1 \\ 0 & 0 \end{bmatrix} \cdot \begin{bmatrix} 1 & 1 \\ 0 & 0 \end{bmatrix} = \begin{bmatrix} 0 & 1 \\ 0 & 0 \end{bmatrix} \cdot \begin{bmatrix} 2 & 0 \\ 0 & 0 \end{bmatrix} = \begin{bmatrix} 0 & 0 \\ 0 & 0 \end{bmatrix}$$

$$\alpha \quad \cdot \quad \beta \quad = \quad \alpha \quad \cdot \quad \gamma = 0$$

but clearly $\beta \neq \gamma$.

9 Remark. The cancellation law holds in any field $\mathbb{F}$ because $\alpha \neq 0$

$$\Rightarrow \exists\, \alpha^{-1} \in \mathbb{F} \text{ and } \alpha^{-1}(\alpha\beta) = \alpha^{-1}(\alpha\gamma)$$

$$\Rightarrow (\alpha^{-1}\alpha)\beta = (\alpha^{-1}\alpha)\gamma \qquad \text{(associativity)}$$

$$\Rightarrow \beta = \gamma \qquad (\alpha^{-1}\alpha = 1)$$

10 Remark. We know some rings for which the cancellation law holds: e.g., $\mathbb{Z}$, $\mathbb{R}[s]$, $\mathbb{C}[s]$, $\mathbb{R}_p(s)$, $\mathbb{R}_{p,o}(s)$, $R(0)$, $R_o(0)$, R_u. Such rings are called integral domains or, better yet, entire rings.

11 Fact. $\forall \alpha \in R$, $\alpha \cdot 0 = 0 \cdot \alpha = 0$.

Proof. $\alpha + 0 = \alpha \Rightarrow \alpha \cdot (\alpha + 0) = \alpha \cdot \alpha \Rightarrow \alpha \cdot \alpha + \alpha \cdot 0 = \alpha \cdot \alpha$. Adding $-(\alpha \cdot \alpha)$ to both sides gives $\alpha \cdot 0 = 0$. Repeat the proof but multiply by α on the right: $0 + \alpha = \alpha \Rightarrow (0 + \alpha) \cdot \alpha = \alpha \cdot \alpha$ etc. gives $0 \cdot \alpha = 0$. ∎

12 Fact. $\forall \alpha, \beta \in R$, $(-\alpha)\beta = -(\alpha \cdot \beta) = \alpha \cdot (-\beta)$.

Proof. $0 = 0 \cdot \beta = [\alpha + (-\alpha)] \cdot \beta = \alpha \cdot \beta + (-\alpha) \cdot \beta \Rightarrow -(\alpha \cdot \beta) = (-\alpha) \cdot \beta$
$0 = \alpha \cdot 0 = \alpha \cdot [\beta + (-\beta)] = \alpha \cdot \beta + \alpha \cdot (-\beta) \Rightarrow -(\alpha \cdot \beta) = \alpha \cdot (-\beta)$. ∎

13 Exercise. Show, from the axioms, that in any ring R,

$$\left(\sum_{i=1}^{n} \alpha_i\right) \cdot \left(\sum_{k=1}^{m} \beta_k\right) = \sum_{i=1}^{n} \sum_{k=1}^{m} \alpha_i \beta_k .$$

Appendix B: Matrices with Elements in a Commutative Ring $\mathbb{K}$

1 Let $\mathbb{K}$ be a commutative ring, i.e., in addition to the standard axioms, we have

2 $$pq = qp \quad \forall p,\, q \in \mathbb{K}.$$

3 Examples. $\mathbb{K}$ might be

(1) any field: $\mathbb{R}$, $\mathbb{R}(s)$, $\cdots$;
(2) $\mathbb{R}[s]$, $\mathbb{R}_p(s)$, $\mathbb{R}_{p,o}(s)$, $\mathcal{R}(0)$, $\mathcal{R}_0(0)$, $\cdots$;
(3) scalar convolution operators: $p*q = q*p$.

4 Fact. For $n > 1$, $\mathbb{K}^{n\times n}$ is a noncommutative ring.

Indeed: $p \in \mathbb{K}^{n\times n}$ means that $\forall i,j \in \underline{n}$, $p_{ij} \in \mathbb{K}$; and

5 $$(P+Q)_{ij} = p_{ij} + q_{ij};\ (PQ)_{ij} = \sum_k p_{ik}q_{kj}.$$

Since all these operations are in the ring $\mathbb{K}$, it is easy to verify that all the axioms of rings are satisfied by P and Q.

6 Fact. The definition and properties of determinants hold as in the case of elements in a field (as long as one does not take inverses!).

For example, if $P,Q \in \mathbb{K}^{n\times n}$, then $\det(PQ) = \det(P)\cdot\det(Q)$.

7 Fact [Cramer's rule]. Let $P \in \mathbb{K}^{n\times n}$, hence $\det P \in \mathbb{K}$. Let Adj(P) denote, as usual, the "classical adjoint" of P. By direct calculation we have

8 (a) $\mathrm{Adj}(P)\,P = P\,\mathrm{Adj}(P) = (\det P)I_n$

(b) $P \in \mathbb{K}^{n\times n}$ has an inverse in $\mathbb{K}^{n\times n}$

$\Leftrightarrow$

9 det P has an inverse in $\mathbb{K}$.

In that case,

10 $$P^{-1} = \text{Adj}(P)\,[\det(P)]^{-1} \in \mathbb{K}^{n\times n}.$$

11 Comment. From (8) and (10) it follows that P has a right inverse iff it has a left inverse: the common right and left inverse of P is called the inverse of P.

12 Proof of Fact 7: indications

(a) (8) is equivalent to [Macl.1]

$$\sum_{k=1}^{n} c_{ki}\, p_{kj} = \sum_{k=1}^{n} p_{ik}\, c_{jk} = \delta_{ij}\,[\det(P)] \quad \forall i,j \in \underline{n},$$

where (1) $\delta_{ij} = 0 \;\; \forall i \neq j$ and $\delta_{ij} = 1 \;\; \forall i = j$.

(2) c_{ij} is the cofactor of element p_{ij} of P, i.e., $c_{ij} = (-1)^{i+j} m_{ij}$ with m_{ij} the minor of order n - 1 obtained by crossing out row i and column j of P.

(b) If P has an inverse $P^{-1} \in \mathbb{K}^{n\times n}$, then by the axioms of $\mathbb{K}$, $\det(P^{-1}) \in \mathbb{K}$; now $P\,P^{-1} = I_n$ implies $[\det(P)]\,[\det(P^{-1})] = 1$: hence (9) holds. Conversely, if (9) holds, then the RHS of (10) $\in \mathbb{K}^{n\times n}$ and is the inverse of P according to (8). ∎

15 Fact. Let $P \in \mathbb{K}^{m\times n}$, $Q \in \mathbb{K}^{n\times m}$; then

$$PQ \in \mathbb{K}^{m\times m} \quad \text{and} \quad QP \in \mathbb{K}^{n\times n}$$

are well defined by the usual operations (see (5) above).

Here is an extremely useful theorem.

16 Theorem. Let $\mathbb{K}$ be a commutative ring. Let $P \in \mathbb{K}^{n_o\times n_i}$ and $F \in \mathbb{K}^{n_i\times n_o}$. Then,

17 (i) $(I_{n_o} + PF)^{-1} \in \mathbb{K}^{n_o\times n_o} \Leftrightarrow (I_{n_i} + FP)^{-1} \in \mathbb{K}^{n_i\times n_i}$.

18 (ii) $(I_{n_o} + PF)^{-1} = I_{n_o} - P(I_{n_i} + FP)^{-1}F$

19 $(I_{n_i} + FP)^{-1} = I_{n_i} - F(I_{n_o} + PF)^{-1}P$.

20 (iii) Either $(I_{n_o} + PF)^{-1} \in \mathbb{K}^{n_o \times n_o}$ or $(I_{n_i} + FP)^{-1} \in \mathbb{K}^{n_i \times n_i}$

implies

21 $$P(I_{n_i} + FP)^{-1} = (I_{n_o} + PF)^{-1}P \in \mathbb{K}^{n_o \times n_o}.$$

22 (iv) $\det(I_{n_i} + FP) = \det(I_{n_o} + PF)$. ∎

23 <u>Remark</u>: <u>Special case of (ii)</u>. If $F = I_n$ and $P \in \mathbb{K}^{n\times n}$, then

24 $$(I_n + P)^{-1} = I_n - (I_n + P)^{-1}P = I_n - P(I_n + P)^{-1}.$$

25 <u>Exercise</u>. Consider the feedback system of Fig. 1, where $G_1 \in \mathbb{R}_p(s)^{n_i \times n_o}$, $G_2 \in \mathbb{R}_p(s)^{n_o \times n_i}$. Calculate the four closed-loop transfer functions $H_{y_j u_i}: u_i \mapsto y_j$. Prove that these four $H_{y_j u_i} \in E(\mathbb{R}_p(s)) \Leftrightarrow \det[I + G_1G_2]$ has an inverse in $\mathbb{R}_p(s)$.

28 <u>Proof of Theorem 16</u>: (a) we handle (i)-(ii) together. Implication ⇒ of (17) and (19) are handled as follows. Assume $(I_{n_o} + PF)^{-1} \in \mathbb{K}^{n_o \times n_o}$. Then

29 $$I_{n_i} - F(I_{n_o} + PF)^{-1}P \in \mathbb{K}^{n_i \times n_i}.$$

We now verify that (29) is the inverse of $(I_{n_i} + FP)$. By calculation in $\mathbb{K}$:

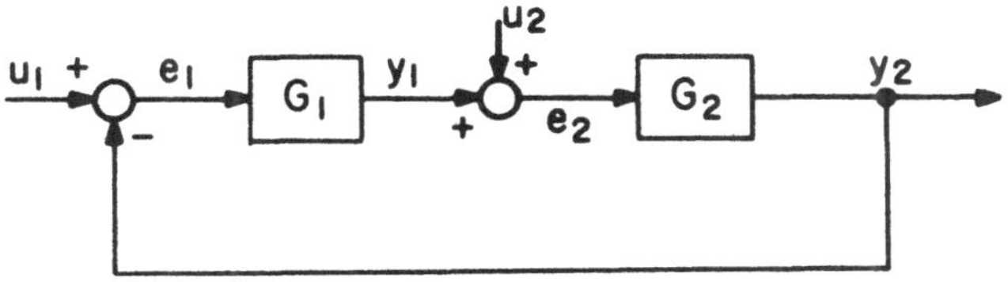

Fig. 1

$$(I_{n_i} + FP)\,[I_{n_i} - F(I_{n_o} + PF)^{-1}P]$$

$$= I_{n_i} + FP - (I_{n_i} + FP)F\,(I_{n_o} + PF)^{-1}P \qquad \text{(distrib. law)}$$

$$= I_{n_i} + FP - F(I_{n_o} + PF)\,(I_{n_o} + PF)^{-1}P \qquad \text{(distrib. law)}$$

$$= I_{n_i}.$$

Hence (19) holds and $(I_{n_i} + FP)^{-1} \in \mathbb{K}^{n_i \times n_i}$ because of (29) and (19).

The implication $\Leftarrow$ of (17) and (18) are handled similarly.

(b) (iii) is established as follows. The distributive law gives

$$(I_{n_o} + PF)P = P + PFP = P(I_{n_i} + FP).$$

Premultiply by $(I_{n_o} + PF)^{-1}$ and postmultiply by $(I_{n_i} + FP)^{-1}$ to obtain equation (21).

(c) We prove (iv). Let

$$M := \left[\begin{array}{c|c} I_{n_i} & F \\ \hline -P & I_{n_o} \end{array}\right] \in \mathbb{K}^{(n_i+n_o)\,(n_i+n_o)}.$$

Now perform on M first $\rho_2 \leftarrow \rho_2 + P\rho_1$, whence M is transformed into

$$M_1 := \left[\begin{array}{c|c} I_{n_i} & F \\ \hline 0 & I_{n_o} + PF \end{array}\right].$$

Perform next on M $\gamma_1 \leftarrow \gamma_1 + \gamma_2\,P$, whence M is transformed into

$$M_2 := \left[\begin{array}{c|c} I_{n_i} + FP & F \\ \hline 0 & I_{n_o} \end{array}\right].$$

Obviously, det M = det M_1 = det M_2, which implies (22). ∎

30 <u>Comment</u>. Appropriate references for Appendices A and B are [Sig.1], [MacL.1], [Jac.1].

Appendix C: Division of a Polynomial Vector on the Left by a Polynomial Matrix

1 Data. Let $D \in \mathbb{R}[s]^{n\times n}$ be a nonsingular polynomial matrix. Let $n \in \mathbb{R}[s]^n$ be a polynomial vector. ∎

By Theorem 2.4.3.37.L we know that there exist unique polynomial vectors q and r in $\mathbb{R}[s]^n$ s.t.

2 $$n = Dq + r \quad \text{with} \quad D^{-1} r \in \mathbb{R}_{p,o}(s)^n.$$

We wish to develop an algorithm which computes q and r. We use ideas of [Emr.1], [Fuh.1], and [Kal.1].

Obviously, the division on the right of a row vector $n \in (\mathbb{R}[s]^n)^T$ by a nonsingular polynomial matrix $D \in \mathbb{R}[s]^{n\times n}$ is similar.

3 Lemma. In the division (2), without loss of generality, we may assume in (1) that D is row reduced with positive row degrees (equiv. D^{-1} is strictly proper), and with highest row-degree coefficient matrix $D_h = I$.

Proof. Indeed, when D is not as in the statement of Lemma 3, consider the following algorithm.

1). Find a unimodular matrix $L(s) \in \mathbb{R}[s]^{n\times n}$, s.t.

4 $$[\tilde{D} \mid \tilde{n}] := L[D \mid n],$$

where $\tilde{D}$ is row reduced with row degrees in decreasing order (the latter can be obtained by reordering rows). Store $L(s)^{-1} \in \mathbb{R}[s]^{n\times n}$.

2). Using the highest row-degree coefficient matrix $\tilde{D}_h$ of $\tilde{D}$, obtain

5 $$[\bar{D} \mid \bar{n}] := [\tilde{D}\, \tilde{D}_h^{-1} \mid \tilde{n}],$$

where $\bar{D}$ is now row reduced with $\bar{D}_h = I$; in particular, $\bar{D}$ can be partitioned into

$$\bar{D} = \left[\begin{array}{c:c} \bar{D}_{11} & \bar{D}_{12} \\ \hdashline 0 & I \end{array}\right] \left.\begin{array}{c} \\ \end{array}\right\} \nu \in \mathbb{R}[s]^{n \times n},$$
$$\underbrace{}_{\nu}$$

where

7   $\bar{D}_{11}$ is row reduced with <u>positive</u> row degrees (equiv. $\bar{D}_{11}^{-1}$ is <u>strictly</u> proper),
and highest row-degree coefficient matrix $\bar{D}_{11h} = I$.

3). Using partitions as in (6), let

8
$$\bar{n} = \left[\begin{array}{c} \bar{n}_1 \\ \hline \bar{n}_2 \end{array}\right] \in \mathbb{R}[s]^n.$$

4). Find the quotient $\bar{q}_1 \in \mathbb{R}[s]^{\nu}$ and remainder $\bar{r}_1 \in \mathbb{R}[s]^{\nu}$ of the division
on the left of $\bar{n}_1 - \bar{D}_{12}\,\bar{n}_2 \in \mathbb{R}[s]^{\nu}$ by $\bar{D}_{11}$, equiv.

9   $\bar{n}_1 - \bar{D}_{12}\,\bar{n}_2 = \bar{D}_{11}\,\bar{q}_1 + \bar{r}_1 \quad$ s.t. $\quad \bar{D}_{11}^{-1}\,\bar{r}_1 \in \mathbb{R}_{p,o}(s)^{\nu}.$

5). Set

10
$$r := L^{-1}\left[\begin{array}{c} r_1 \\ \hline \theta \end{array}\right] \in \mathbb{R}[s]^n \quad \text{and} \quad q := \tilde{D}_h^{-1}\left[\begin{array}{c} \bar{q}_1 \\ \hline \bar{n}_2 \end{array}\right] \in \mathbb{R}[s]^r.$$

                                                                End of Algo.

<u>Claim.</u>   The polynomial vectors r and q, given by (10), are the remainder
and quotient of the division on the left, (2), of n by D.

        Indeed, by (6)-(10),

11                    $\bar{n} = \bar{D}\,\bar{q} + \bar{r} \quad$ with $\quad \bar{D}^{-1}\,\bar{r} \in \mathbb{R}_{p,o}(s)^n,$

where

12
$$\bar{q} := \left[\begin{array}{c} \bar{q}_1 \\ \hline \bar{n}_2 \end{array}\right] \in \mathbb{R}[s]^n \quad \text{and} \quad \bar{r} := \left[\begin{array}{c} \bar{r}_1 \\ \hline \theta \end{array}\right] \in \mathbb{R}[s]^n.$$

Hence by (4)-(5) and (10)-(12),

2 $$n = Dq + r \quad \text{with} \quad D^{-1} r \in \mathbb{R}_{p,o}(s)^n .$$

Therefore, the claim holds and Lemma 3 must hold. Indeed, in the algorithm, division on the left by D, (2), is reduced to division on the left by $\bar{D}_{11}$, (9), where, (7), $\bar{D}_{11}$ has the properties requested in the statement of Lemma 3. ■

13 Comment. In the algorithm above it may happen that $\bar{D}^{-1}$ is strictly proper. Then in (6), $\bar{D} = \bar{D}_{11}$ and Lemma 3 still holds with obvious modifications to the algorithm.

By Lemma 3, without loss of generality, the division on the left of a polynomial vector $n \in \mathbb{R}[s]^n$ by a nonsingular polynomial matrix can be studied under the following data.

15 Data. We are given a polynomial vector $n \in \mathbb{R}[s]^n$ and a nonsingular matrix $D \in \mathbb{R}[s]^{n \times n}$ which is row reduced with positive row degrees $\rho_i > 0$, $\forall i \in \underline{n}$, and with highest row-degree coefficient matrix $D_h = I$. Hence

16 $$D(s) = \operatorname{diag}[s^{\rho_i}] + D_\ell(s),$$

where

17 $$\rho_i := \partial_{ri}[D] > 0, \quad \forall i \in \underline{n},$$

and

18 $D_\ell(s) \in \mathbb{R}[s]^{n \times n}$ accounts for the terms of degree smaller than ρ_i for every entry in each row of D. ■

Before giving an algorithm for division on the left, we should observe the following.

20 Lemma. Let $D \in \mathbb{R}[s]^{n \times n}$ be as in (15) and let the set $\mathbb{R}_D$ be defined by

21 $$\mathbb{R}_D := \{x \in \mathbb{R}[s]^n : D^{-1} x \in \mathbb{R}_{p,o}(s)^n\}.$$

U.t.c.

22 (a) $\mathbb{R}_D$ is an $\mathbb{R}$-linear space;

(b) $x \in \mathbb{R}_D$

if and only if

23 $$x(s) = (x_i(s))^T_{i\in\underline{n}},$$

where

$$x_i(s) := x_{i1}\, s^{\rho_i - 1} + x_{i2}\, s^{\rho_i - 2} + \cdots + x_{i\rho_i} \in \mathbb{R}[s], \quad \forall i \in \underline{n};$$

$$\rho_i = \partial_{ri}[D], \quad \forall i \in \underline{n};$$

some of the x_{ij}'s, $j \in \rho_i$, may be zero.

Proof. Exercise. ∎

24 Exercise. Use contradiction to prove that the division (2) leads to a unique quotient and a unique remainder. ∎

Let us now return to the division on the left by $D \in \mathbb{R}[s]^{n\times n}$, where, $\forall x \in \mathbb{R}[s]^n$, there exists a unique quotient q and remainder r s.t.

$$x = Dq + r \quad \text{s.t.} \quad D^{-1}\, r \in \mathbb{R}_{p,o}(s)^n.$$

Hence we can define a quotient and remainder operator $\tilde{q}$, resp. $\tilde{r}$ s.t.

25 $$\tilde{q} : \mathbb{R}[s]^n \to \mathbb{R}[s]^n : x \mapsto \tilde{q}(x),$$

26 $$\tilde{r} : \mathbb{R}[s]^n \to \mathbb{R}_D \quad : x \mapsto \tilde{r}(x),$$

where

27 $$x = D\,\tilde{q}(x) + \tilde{r}(x).$$

28 Exercise. Show that the operators $\tilde{q}$ and $\tilde{r}$ are s.t.,

29 (a) $\tilde{q}$ and $\tilde{r}$ are $\mathbb{R}$-linear operators,

(b) $\forall x(s) \in \mathbb{R}[s]^n$,

30 $$\tilde{r}(sx(s)) = \tilde{r}\big(s\tilde{r}(x(s))\big).$$ ∎

We study now a basic operation for division on the left by D.

35 Lemma. Let D be as in (15). Let $x(s) \in \mathbb{R}_D$ (equiv. x(s) is given by (23)). U.t.c.

36 $$sx(s) = D(s)\, \tilde{q}(sx(s)) + \tilde{r}(sx(s)),$$

where, with (23),

37 $$\tilde{q}(sx(s)) = (x_{i1})^T_{i \in \underline{n}} \in \mathbb{R}^n$$

and

38 $$D(s)^{-1}\, \tilde{r}(sx(s)) \in \mathbb{R}_{p,o}(s)^n \text{ (equiv. } \tilde{r}(sx(s)) \in \mathbb{R}_D).$$

Proof. Clearly from (36)-(37), (23), and (16)-(18),

39 $$\tilde{r}(sx(s)) = sx(s) - D(s)\, \tilde{q}(sx(s)) \in \mathbb{R}[s]^n$$

is s.t., $\forall i \in \underline{n}$, the ith component reads

40 $$x_{i2}\, s^{\rho_i - 1} + x_{i3}\, s^{\rho_i - 2} + \cdots + x_{i\rho_i}\, s - D_{\ell i}(s)\, \tilde{q}(sx(s)),$$

where $D_{\ell i}$ denotes the ith row of $D_\ell \in \mathbb{R}[s]^{n \times n}$ in (16). Clearly, $\forall i \in \underline{n}$, the degree of the polynomial (40) is smaller than ρ_i. Therefore, since D is row reduced as in (15), (38) must hold. ∎

41 Comment. For D as in (15) and for $x \in \mathbb{R}_D$, Lemma 35 shows how to perform the division on the left of sx(s) by D: $\forall i \in \underline{n}$, the ith component of the (constant) quotient vector is the coefficient of s^{ρ_i} of the ith (polynomial) component of sx(s).

We are now able to prove the main theorem of this appendix; its statement contains the division algorithm.

45 Theorem [Division on the left of a polynomial vector n by a polynomial matrix D]. Let $D \in \mathbb{R}[s]^{n \times n}$ be as in (15) and let $n \in \mathbb{R}[s]^n$ be given by

46 $$n(s) = n_o\, s^k + n_1\, s^{k-1} + \cdots + n_k,$$

where $n_t \in \mathbb{R}^n$, $\forall t = 0, 1, \cdots, k$.

Consider the quotient and remainder operators $\tilde{q}$, resp. $\tilde{r}$, as defined in (25)-(27). Now consider the discrete-time system Σ described by the equations

47a $$x_{t+1}(s) = \tilde{r}(sx_t(s)) + n_t,$$

$$\forall t = 0, 1, 2, \cdots, k,$$

47b $$y_{t+1} = \tilde{q}(sx_t(s)),$$

where, $\forall t = 0, 1, \cdots, k$, $n_t \in \mathbb{R}^n$ is the coefficient vector of s^{k-t} in (46), and

48 $$x_o(s) := \theta \in \mathbb{R}_D.$$

U.t.c.

49 (a) $\forall t = 1, 2, \cdots, k+1$, $x_t(s) \in \mathbb{R}_D$ and $y_t \in \mathbb{R}^n$.

(b) The remainder and quotient of the division on the left of n by D are given by

50 $$\tilde{r}(n(s)) = x_{k+1}(s) \in \mathbb{R}_D,$$

51 $$\tilde{q}(n(s)) = \sum_{t=1}^{k} y_{t+1}\, s^{k-t} \in \mathbb{R}[s]^n.$$

52 Comments. (a) The remainder is the state of Σ at $t = k + 1$ and the quotient is given by the outputs y_t for $t = 2, \cdots, k+1$.

(b) At each iteration we divide $sx(s)$ by D with $x \in \mathbb{R}_D$.

53 Proof of Theorem 45

(a): This follows from (47)-(48) and Lemma 35. Note with care that $\mathbb{R}_D$ is an $\mathbb{R}$-linear space and $n_t \in \mathbb{R}_D$.

(b) (50) is established by induction, using (29)-(30).

Indeed:

$$x_1(s) = n_o = \tilde{r}(n_o)$$
$$x_2(s) = \tilde{r}(s\tilde{r}(n_o)) + n_1 = \tilde{r}(sn_o) + \tilde{r}(n_1) = \tilde{r}(sn_o + n_1)$$
$$x_3(s) = \tilde{r}(s\tilde{r}(sn_o + n_1)) + n_2 = \tilde{r}(s^2 n_o + sn_1 + n_2)$$
$$\cdots\cdots$$
$$x_{k+1}(s) = \tilde{r}(s^k n_o + s^{k-1} n_1 + \cdots + n_k) = \tilde{r}(n(s)).$$

(51) follows also by induction.

Indeed:

$$sn_o + n_1 = D(s)\,\tilde{q}(sn_o) + \tilde{r}(sn_o + n_1)$$

$$= D(s)\,y_2 + x_2(s)$$

$$s^2n_o + sn_1 + n_2 = s(sn_o + n_1) + n_2$$

$$= s(D(s)\,y_2 + x_2(s)) + n_2$$

$$= sD(s)\,y_2 + sx_2(s) + n_2$$

$$= D(s)(sy_2 + y_3) + x_3(s)$$

...

$$n(s) = s^k\,n_o + s^{k-1}\,n_1 + \cdots + n_k$$

$$= D(s)\,\Big(\sum_{t=1}^{k} y_{t+1}\,s^{k-t}\Big) + x_{k+1}(s)$$

$$= D(s)\,\tilde{q}(n(s)) + \tilde{r}(n(s)).$$ ■

55 Example. Let

$$D(s) = \begin{bmatrix} (s+1)^2(s+2) & s+2 \\ 0 & (s+2)^2 \end{bmatrix}, \quad n(s) = \begin{bmatrix} s^4 \\ s^2 \end{bmatrix}.$$

Then

$$\tilde{q}(n(s)) = \begin{bmatrix} s-4 \\ 1 \end{bmatrix} \text{ and } \tilde{r}(n(s)) = \begin{bmatrix} 11s^2 + 17s + 6 \\ -4s-4 \end{bmatrix}.$$ ■

60 Final Comment: Division of a Polynomial Matrix $N \in \mathbb{R}[s]^{n \times p}$ on the Left by a Nonsingular Polynomial Matrix $D \in \mathbb{R}[s]^{n \times n}$. This division is carried out by successively dividing on the left the columns of N by D.

References

Cal. 1 F. M. Callier and C. A. Desoer, "Stabilization, Tracking and Disturbance Rejection in Multivariable Convolution Systems," Annales de la Société Scientifique de Bruxelles, T. 94, I, pp. 7-51, 1980.

Cal. 2 F. M. Callier, V. H. L. Cheng, and C. A. Desoer, "Dynamic Interpretation of Poles and Transmission Zeros for Distributed Multivariable Systems," IEEE Trans. Circ. and Syst., Vol. CAS-28, pp. 300-307, 1981.

Che. 1 C. T. Chen, "Introduction to Linear System Theory," Holt, Rinehart and Winston, New York, 1970.

Chen. 1 M. J. Chen, C. A. Desoer, and G. F. Franklin, "Algorithmic Design of Single-Input Single-Output Systems with a Two-Input One-Output Controller," Memorandum Electronics Research Laboratory, University of California, Berkeley, No. UCB/ERL M81/24, July 1921.

Cru. 1 J. B. Cruz, J. S. Freudenberg, and D. P. Looze, "A Relation between Sensitivity and Stability of Multivariable Feedback Systems," IEEE Trans. Auto. Control, Vol. AC-26, pp. 66-74, February 1981.

Dav. 1 E. J. Davison, "The Robust Control of a Servomechanism Problem for Linear Time-Invariant Multivariable Systems," IEEE Trans. Auto. Control, Vol. AC-21, pp. 25-34, February 1976.

Des. 1 C. A. Desoer and J. D. Schulman, "Zeros and Poles of Matrix Transfer Functions and their Dynamical Interpretations," IEEE Trans. Circ. and Syst., Vol. CAS-21, January 1974.

Des. 2 C. A. Desoer and Y. T. Wang, "Linear Time-Invariant Robust Servomechanism Problem: A Self-Contained Exposition," in "Advances in Control and Dynamical Systems," Vol. 16, C. T. Leondes (Ed.), Academic Press, New York, pp. 81-129, 1980.

Des. 3 C. A. Desoer and Y. T. Wang, "Foundations of Feedback Theory for Nonlinear Dynamical Systems," IEEE Trans. Circ. and Syst., Vol. CAS-27, pp. 320-323, April 1980.

Des. 4 C. A. Desoer and M. Vidyasagar, "Feedback Systems: Input-Output Properties," Academic Press, New York, 1975.

Des. 5 C. A. Desoer and M. J. Chen, "Design of Multivariable Feedback Systems with Stable Plants," IEEE Trans. Auto. Control, Vol. AC-26, 2, pp. 408-415, April 1981.

Doy. 1 J. C. Doyle and G. Stein, "Multivariable Feedback Design: Concepts for a Classical/Modern Synthesis," IEEE Trans. Auto. Control, Vol. AC-26, pp. 4-17, February 1981.

Emr. 1 E. Emre, "The Polynomial Equation $QQ_c + RP_c = \Phi$ with Application to Dynamic Feedback," SIAM Jour. Contr. and Opt., Vol. 18, pp. 611-620, November 1980.

Fra. 1 B. A. Francis and M. Vidyasagar, "Algebraic and Topological Aspects of the Servo Problem for Lumped Linear Systems," S & IS Report No. 8003, Yale University, New Haven, Conn., 1980.

Fuh. 1 P. A. Fuhrmann, "Algebraic System Theory: An Analyst's Point of View," Jour. Franklin Inst., Vol. 301, pp. 521-540, 1976.

Gan. 1 F. R. Gantmacher, "The Theory of Matrices," Vol. 1, Chelsea, New York, 1959.

Gar. 1 B. S. Garbow, M. M. Boyle, J. J. Dongarra, and C. B. Moler, "Matrix Eigensystem Routines -- EISPACK Guide Extension," Lecture Notes in Computer Science, No. 51, Springer Verlag, New York, 1977.

Goh. 1 I. Gohberg and L. Rodman, "On Spectral Analysis of Non-Monic Matrix and Operator Polynomials, I. Reduction to Monic Polynomials," Israel Jour. Math., Vol. 30, pp. 133-151, 1978.

Goh. 2 I. Gohberg and L. Rodman, "On Spectral Analysis of Non-monic Matrix and Operator Polynomials, II. Dependence on the Finite Spectral Data," Israel Jour. Math., Vol. 30, No. 4, pp. 321-334, 1978.

Hun. 1 N. T. Hung and B. D. O. Anderson, "Triangularization for the Design of Multivariable Control Systems," IEEE Trans. Auto. Contr., Vol. AC-24, pp. 455-460, 1979.

Jac. 1 N. Jacobson, "Basic Algebra I," Freeman, San Francisco, 1974.

Kai. 1 T. Kailath, "Linear Systems," Prentice-Hall, Englewood Cliffs, N. J., 1980.

Kal. 1 R. E. Kalman, P. L. Falb, and M. A. Arbib, "Topics in Mathematical System Theory," McGraw-Hill, New York, 1969.

Kuc. 1 V. Kucera, "Discrete Linear Control, The Polynomial Approach," John Wiley, New York, 1979.

MacD. 1 C. C. MacDuffee, "The Theory of Matrices," Springer-Verlag, Berlin, 1933, reprinted Chelsea, New York, 1956.

MacL. 1 S. MacLane and G. Birkhoff, "Algebra," 2nd. ed., Mcmillan, New York, 1979 (1st. ed., 1967).

Nob. 1 B. Noble and J. W. Daniel, "Applied Linear Algebra," Prentice-Hall, Englewood Cliffs, N. J., 1977, rev. ed. (1st. ed., 1969).

Per. 1 L. Pernebo, "An Algebraic Theory for the Design of Controllers of Linear Multivariable Systems, Part I: Structure Matrices and Feedforward Design, Part II: Feedback Realizations and Feedback Design," IEEE Trans. Auto. Control, Vol. AC-26, pp. 171-193, February 1981.

Pol. 1 E. Polak, "Optimization-based Computer-aided Design of Control Systems," Proceedings of the 1981 Joint Automatic Control Conference.

Pug. 1 A. C. Pugh and P. R. Ratcliffe, "On the Zeros and Poles of a Rational Matrix," Int. Jour. Control, Vol. 30, pp. 213-226, 1979.

Ros. 1 H. H. Rosenbrock, "State Space and Multivariable Theory," Wiley, New York, 1970.

Ros. 2 H. H. Rosenbrock and G. E. Hayton, "The General Problem of Pole Assignment," Int. Jour. Control, Vol. 27, No. 6, pp. 837-852, 1978.

Rud. 1 W. Rudin, "Real and Complex Analysis," McGraw-Hill, New York, 1974.

Saf. 1 M. G. Safonov, A. J. Laub, and G. L. Hartmann, "Feedback Properties of Multivariable Systems: The Role and Use of the Return Difference Matrix," IEEE Trans. Auto. Control, Vol. AC-26, pp. 47-65, February 1981.

Saf. 2 M. G. Safonov, "Stability and Robustness of Multivariable Feedback Systems," MIT Press, Cambridge, Mass., 1980.

Sig. 1 L. E. Sigler, "Algebra," Undergraduate Texts in Mathematics, Springer Verlag, New York, 1976.

Ste. 1 G. W. Stewart, "Introduction to Matrix Computations," Academic Press, New York, 1973.

Tem. 1 G. C. Temes and J. W. La Patra, "Circuit Synthesis and Design," McGraw-Hill, New York, 1977.

Ver. 1 G. Verghese, "Infinite-Frequency Behavior in Generalized Dynamical Systems," Ph.D. Thesis, Stanford University, Stanford, Calif., December 1978.

Vid. 1 M. Vidyasagar, H. Schneider, and B. A. Francis, "Algebraic and Topological Aspects of Feedback Stabilization," Tech., Report No. 80-09, Dept. of Electrical Engineering, University of Waterloo, September 1980.

Vid. 2 M. Vidyasagar, "Nonlinear System Analysis," Prentice-Hall, Englewood Cliffs, N. J., 1978.

Vid. 3 M. Vidyasagar, "Input-Output Analysis of Large-Scale Interconnected Systems," Lecture Notes in Control and Information Sciences, Springer Verlag, New York, 1981.

Wil. 1 J. C. Willems, "The Analysis of Feedback Systems," MIT Press, Cambridge, Mass, 1971.

You. 1 D. C. Youla, J. J. Bongiorno, Jr., and C. N. Lu, "Single-Loop Feedback Stabilization of Linear Multivariable Dynamical Plants," Automatica, Vol. 10, pp. 159-173, 1974.

Zam. 1 G. Zames, "Feedback and Optimal Sensitivity: Model Reference Transformations, Multiplicative Seminorms and Approximate Inverses," IEEE Trans. Auto. Control, Vol. AC-26, pp. 301-320, April 1981.

Symbols

I. Sets

$\in$	$a \in A$	a is an element of A; a belongs to A
$\subset$	$A \subset B$	set A is contained in set B; A is a subset of B
$\cup$	$A \cup B$	union of set A with set B
$\cap$	$A \cap B$	intersection of set A and set B
$\Rightarrow$	$p \Rightarrow q$	p implies q; equivalently, "not q" implies "not p"
$\Leftarrow$	$p \Leftarrow q$	q implies p
$\Leftrightarrow$	$p \Leftrightarrow q$	p if and only if q; equivalently, p implies q and q implies p
	$\mathring{M}$	interior of the set M
	$\bar{M}$	closure of the set M (Note: For $z \in \mathbb{C}$, $\bar{z}$ denotes the complex conjugate of z)
	M^c	complement of the set M
:=	$A := B$	the set A is by definition the set B
=:	$A =: B$	the set B is by definition the set A
\	$A \setminus B$	the set difference of set A minus set B, equiv. $A \cap B^c$
$\times$	$A \times B$	the cartesian product of set A times set B; equiv. $A \times B = \{(a, b); a \in A, b \in B\}$
	ϕ	the empty set

II. Algebra

$\mathbb{C}$	field of complex numbers
$\underline{k}$	the set of integers $\{1, 2, \cdots, k\}$
$\mathbb{N}$	set of nonnegative integers, namely, $\{0, 1, 2, \cdots\}$
$\mathbb{R}$	field of real numbers
$\mathbb{Z}$	ring of integers, namely, $\{\cdots, -1, 0, 1, 2, \cdots\}$
$\mathbb{Q}$	field of rational numbers
R	a ring
$\mathbb{K}$	a commutative ring
$\mathbb{F}$	a field
$\mathbb{F}[x]$	ring of polynomials in one variable with coefficients in the field $\mathbb{F}$ (e.g., $\mathbb{R}[s]$, $\mathbb{C}[s]$)

$\mathbb{F}(x)$	field of rational functions in one variable with coefficients in $\mathbb{F}$ (e.g., $\mathbb{R}(s)$, $\mathbb{C}(s)$)
$\mathbb{R}_p(s)$, $\mathbb{R}_{p,o}(s)$	the ring of proper (equiv. bounded at infinity), resp. strictly proper (equiv. zero at infinity), rational functions in s with coefficients in $\mathbb{R}$
$\mathbb{C}_+$	$:= \{s \in \mathbb{C} : \mathrm{Re}\, s \geq 0\}$, equiv. the closed right-half of the complex plane
$\mathcal{U}$	an undesirable subset of $\mathbb{C}$, which is symmetric w.r.t. the real axis and contains $\mathbb{C}_+$
$\mathcal{R}(0)$, $\mathcal{R}_o(0)$	the subring of elements of $\mathbb{R}_p(s)$, resp. $\mathbb{R}_{p,o}(s)$, that are analytic in $\mathbb{C}_+$ (equiv. with no poles in $\mathbb{C}_+$)
$\mathcal{R}_{\mathcal{U}}$	the subring of elements of $\mathcal{R}(0)$ that are analytic in $\mathcal{U}$
$(x_k)_{k \in K}$	family of elements; K := index set; the family is a map $K \to X$
A^n	set of n-tuples of elements belonging to the set A (e.g., $\mathbb{R}^n$, $\mathbb{R}[s]^n$, $\mathbb{R}(s)^n$, ...)
$A^{p \times q}$	set of $p \times q$ arrays of elements belonging to the set A; equiv. the set of $p \times q$ matrices with entries in the set A, (e.g., $\mathbb{R}^{p \times q}$, $\mathbb{R}[s]^{p \times q}$, $\mathbb{R}(s)^{p \times q}$, ...)
$A \in \mathcal{E}[B]$	matrix A has entries in the set B
∂a	the degree of the polynomial a
$\partial_{ri}[A]$	the ith row degree of the polynomial matrix A
$\partial_{cj}[A]$	the jth column degree of the polynomial matrix A
$\partial_M[A]$	the Mcmillan degree of the rational matrix A
$p \mid q$	the polynomial p divides the polynomial q without remainder, or equiv. p is a factor of q
$A \sim B$	the polynomial matrix A is equivalent to the polynomial matrix B, or equiv. there exist unimodular polynomial matrices L and R such that $A = LBR$
$a \sim b$	the polynomial a is equivalent to the polynomial b, or equiv. there exists a nonzero constant k s.t. $a = kb$
a^*, A^*	the complex conjugate transpose of the complex vector a, resp. matrix A
a^T, A^T	the transpose of the vector a, resp. matrix A
ρ_i	the ith row of a matrix
γ_j	the jth column of a matrix
$\mathcal{V}$	the vector space $\mathcal{V}$ (also called linear space)
$\mathcal{M}$	the module $\mathcal{M}$

θ	the zero element of a vector space or module
$\mathcal{N}(A)$	the null space of matrix A (or the linear map A)
$\mathcal{R}(A)$	the range or image of the matrix A (or the linear map A)
$rk(A)$	the rank of matrix A
$\sigma(A)$	the spectrum of matrix A (equiv. the set of eigenvalues of A)
$\sigma_{max}[A]$, $\bar{\sigma}[A]$	the largest singular value of matrix A
$\sigma_{min}[A]$	the smallest singular value of matrix A
$\mathcal{V}_1 \oplus \mathcal{V}_2$, $\mathcal{M}_1 \oplus \mathcal{M}_2$	the direct sum of two vector spaces, resp. modules
$\mathcal{V}_1 \perp \mathcal{V}_2$	vector space $\mathcal{V}_1$ is orthogonal to vector space $\mathcal{V}_2$
$\mathcal{V}_1 \overset{\perp}{\oplus} \mathcal{V}_2$	the direct orthogonal sum of vector spaces $\mathcal{V}_1$ and $\mathcal{V}_2$

III. Analysis

$f : A \to B$	f is a function or map, mapping a domain A into the codomain B; also noted $x \mapsto f(x)$, i.e., f is a function which associates with $x \in A$ the image $f(x) \in B$
dom[f]	the domain of the map f
codom[f]	the codomain of the map f
$\mathcal{R}[f]$	the range or image of the map f
$\mathcal{N}[f]$	the nullspace of the map f
$\exists$	there exists
$\exists!$	there exists a unique element
s.t.	such that
$\mathbb{C}_+$	$:= \{s \in \mathbb{C} : \text{Re } s \geq 0\}$
$\mathring{\mathbb{C}}_-$	$:= \{s \in \mathbb{C} : \text{Re } s < 0\}$, equiv. the open left-half of the complex plane
$\mathring{\mathbb{C}}_+$	$:= \{s \in \mathbb{C} : \text{Re } s > 0\}$
p	the differentiation operator, e.g., $pf = df/dt$
C	the space of continuous functions
C^k	the space of k-times continuously differentiable functions
C^∞	the space of infinitely continuously differentiable functions
$\\|f\\|$	the norm of f
$\mathcal{P}[f]$, $\mathcal{P}[H]$	the set of poles of the vector function f, resp. the matrix function H
$\mathcal{Z}[f]$, $\mathcal{Z}[H]$	the set of zeros of the vector function f, resp. the matrix function H

$(x_\lambda)_{\lambda\in L}$	the family of elements x_λ with index set L, (special case: sequences: $(x_\lambda)_{\lambda\in\mathbb{N}}$ is often written $(x_\lambda)_0^\infty$ or simply (x_λ)).
$\hat{f}$, $\hat{H}$	the Laplace transform of the vector function f, resp. matrix function H (also denoted by $\mathcal{L}[f]$, resp. $\mathcal{L}[H]$)
$\mathcal{L}^{-1}[\hat{f}]$, $\mathcal{L}^{-1}[\hat{H}]$	the inverse Laplace transform of the vector function $\hat{f}$, resp. matrix function $\hat{H}$.

List of Abbreviations

PMD	polynomial matrix system description
equiv.	equivalently
s.v.d.	singular value decomposition
s.t.	such that
exp. stable	exponentially stable
I/O map	input-output map
r.d.	right divisor
ℓ.d.	left divisor
ℓ.m.	left multiple
r.m.	right multiple
g.c.r.d.	greatest common right divisor
g.c.ℓ.d.	greatest common left divisor
c.r.d.	common right divisor
c.ℓ.d.	common left divisor
r.c.	right coprime
ℓ.c.	left coprime
e.o.	elementary operation
e.r.o.	elementary row operation
e.c.o.	elementary column operation
e.m.	elementary matrix
ℓ.e.m.	left elementary matrix
r.e.m.	right elementary matrix
r.f.	right fraction
r.c.f.	right coprime fraction
ℓ.f.	left fraction
ℓ.c.f.	left coprime fraction
SMM-form	Smith-Mcmillan form
c.r.	column reduced
r.r.	row reduced
p. suff. diff.	piecewise sufficiently differentiable
z-i p-s trajectory	zero-input pseudo-state trajectory
z-i response	zero-input response
z-s p-s trajectory	zero-input pseudo-state trajectory

z-s response	zero-state response
i-d zero	input-decoupling zero
o-d zero	output-decoupling zero
r.c.r.	row-column reduced
r.ℓ.f.	right-left fraction
r.ℓ.c.f.	right-left coprime fraction
int. pr.	internally proper
ℓ.r.f.	left-right fraction
ℓ.r.c.f.	left-right coprime fraction
char. poly.	characteristic polynomial
eqn.	equation
RHS	right-hand side
LHS	left-hand side
comp. eqn.	compensator equation

Index